Cat Detective

www.**booksattransworld**.co.uk

Also by Vicky Halls

Cat Confidential

CAT DETECTIVE

Solving the Mystery of Your Cat's Behaviour

VICKY HALLS

BANTAM PRESS

LONDON · TORONTO · SYDNEY · AUCKLAND · JOHANNESBURG

TRANSWORLD PUBLISHERS
61–63 Uxbridge Road, London W5 5SA
a division of The Random House Group Ltd

RANDOM HOUSE AUSTRALIA (PTY) LTD
20 Alfred Street, Milsons Point, Sydney,
New South Wales 2061, Australia

RANDOM HOUSE NEW ZEALAND LTD
18 Poland Road, Glenfield, Auckland 10, New Zealand

RANDOM HOUSE SOUTH AFRICA (PTY) LTD
Isle of Houghton, Corner Boundary Road and Carse O'Gowrie, Houghton 2198, South Africa

Published 2005 by Bantam Press
a division of Transworld Publishers

A catalogue record for this book is available from the British Library.
ISBN 0593 052773

Typeset in 12/15.75pt Cochin by
Falcon Oast Graphic Art Ltd

Printed in Great Britain by
Mackays of Chatham, Chatham, Kent

1 3 5 7 9 10 8 6 4 2

Papers used by Transworld Publishers are natural, recyclable products made from wood grown
in sustainable forests. The manufacturing processes conform to the environmental
regulations of the country of origin.

This book is dedicated to the memory of
Spooky, Hoppy, Zulu, Bln and Puddy Halls
Jenny Kenyon
Bo-Bo Phillips
Emma Maidment
Oscar Sexton
Jake and Hilda Clinton
Sinbad and Dibley Bullen
Trixie and Honey Harris
Stan and Ollie Walter
Paddington and Tigger Hall
Frodo Lake
Puff Wake-Murrell
Tinker Woolf
Jerry Hannay
Morgan Morgan
Jasper Philip
Mortimer Chilton
and all the other cats who have touched my life and are no more.

CONTENTS

ACKNOWLEDGEMENTS

I could not have carried on with my counselling work, written this book and remained sane without the patience and love of all my fabulous and loyal friends. I am nothing without you! Thank you, Mum, as always for your wisdom and humour and thank you, Peter, for continuing to care for my beautiful Cornish cats.

Thanks to Mary Pachnos, my wonderful agent, for her guidance and friendship. Thanks also to Francesca Liversidge, Emily Furniss and the rest of the team at Transworld who have continued to support me with such enthusiasm and good humour.

INTRODUCTION

I HAVE A VERY UNUSUAL JOB THAT MANY PEOPLE ENVY BUT few understand. I am a cat shrink.

Since the late 1980s I have been working with cats, initially in rescue centres and then in veterinary practice. I have helped cats when they were homeless and I've nursed them when they were sick. Since 1995 I have been fortunate enough to be in a position to care for them emotionally too. With a great deal of support and guidance at the beginning from Dr Peter Neville, a pioneer in the science of cat behaviour counselling, I now spend my days visiting owners whose cats are experiencing a wide variety of behavioural, emotional and psychological problems. It can be a time-consuming and often stressful occupation but the results make the whole process addictive.

For years now I have been helping people who have tolerated bad situations for long periods of time. Their cats' behavioural problems make life difficult for everyone. I often work with owners at the very end of their tether, physically and emotionally. They are unhappy, the cats are unhappy and the process of repair and harmony restoration can be a real struggle. However, whether a Devon Rex is demolishing a carpet, a moggy is ripping its fur out or a Siamese is peeing in the toaster, there is always a common denominator: without exception the owners all remark that they have 'tried everything'. Sources of well-meaning advice are plentiful but, if incorrect or misunderstood, will merely prolong the agony. Wouldn't it be great if there was one single source of good common sense and practical information that could potentially lead to a quicker solution?

I do not profess to know everything about cats; in reality my knowledge makes it apparent how little all of us actually *do* know. My work could be described as the artistic interpretation of the science of cat behaviour counselling and everything described within this book shows my methods, techniques and opinions. There will be many other excellent practitioners who do things differently but still get results. That aside, I can and *do* make a difference, but there are plenty of occasions when the owners can do it all by themselves. I have written *Cat Detective* with that in mind. Ultimately, though, there is no substitute for an experienced cat behaviour counsellor for complicated or severe cases, and I hope this book will also show which situations need help straight away. Many problems result from massive misunderstandings about the species we share our homes with and miracle cures can be effected by making subtle changes to a cat's environment once the problem is truly appreciated in feline terms.

The whole concept of cat behaviour therapy is shrouded in mystery and often ridiculed by those who know nothing about it. They have an image of a bespectacled shrink leaning over a cat reclining upon a couch. The cat is pouring forth tragic tales of its deprived kittenhood peppered with cruelty and neglect. Gently holding the cat's paw, the shrink talks things through and helps the cat come to terms with its past. Whilst I love the image (oh, that it were quite so easy), it is a fantasy. So, by the way, is the image of the owner on the couch pouring forth his or her tragic tales of a deprived childhood peppered with cruelty and neglect. I won't deny this can happen on occasion (I think some people get a little confused about my qualifications) but ultimately I'm there to help the cat.

Cats are enigmatic creatures with subtle and ambiguous elements to their characters. Any attempts to manipulate their behaviour by direct intervention or enforced training are doomed to failure. I do not work directly with my patients; there have even been times when they have remained steadfastly under the bed and all I've seen is a pair of saucer eyes. It really doesn't matter because I work with the owners. The environment and how the owner interacts with the cat are critical and even slight changes can dramatically alter the cat's behaviour. I have often been accused of witchery as the cat improves immediately after my visit without anything apparently being done. The trick is that something has been done. The owner has started to understand why the cat is soiling (for example) and he or she is displaying a less hostile body language in the presence of the offender. That is sometimes sufficient to take the heat off and reduce stress levels. Never underestimate the benefits of positive thought and relaxation when your cat is behaving badly.

Observers of my profession often remark that all pet

problems are caused by the owners. It is certainly true that, without exception, owners all have an enormous burden of guilt by the time I get to see them, but it's often unfounded. Cat problems are usually a cat thing so let's not over-emphasize the role we play in all this. Humans compound or complicate issues but it is always necessary to look at things from the feline perspective before presuming the fault lies solely with the owner. If we as a species are guilty of anything it is irresponsible over-breeding, owning cats in situations that represent over-crowding, and lack of knowledge about the nature of these complicated creatures.

A significant percentage of problems are, however, purely relationship issues between cat and owner. These are difficult to resolve on a Do It Yourself basis because it is impossible to distance yourself sufficiently from the problem, when you are part of it, to see the underlying cause. I have attempted to discuss the owner/cat relationship in Chapter 10 but often, if a person has a rather dysfunctional relationship with their pet, they are the last to recognize it. I cannot force people to review 'what's in it for the cat' but I hope that my advice may help some owners who genuinely feel that their cats are not as happy as they could be despite all the love and attention.

In Chapter 1 I have spent some time discussing what actually constitutes a behavioural problem. This may seem like stating the obvious but it can be a very personal thing; after all one man's problem is another man's idiosyncrasy. One lady, for example, telephoned me complaining that her cat loved her too much and another felt that her cat's desire to eat with his paw merited intervention. Most owners, however, are concerned about the more common problems such as house soiling and aggression, and these are what I aim to cover in this book. You just need to know where to look. Chapter 7, for instance, is devoted to inter-cat aggression and the whole concept of multi-cat households, and is a must for all people with (or thinking of acquiring) two or more cats. This was always going to be the longest chapter in the book since I consider multi-cat households to be the cause of many potential problems. I have also encouraged people to assess whether their cats are actually as happy as they might think. This may sound like a pointless exercise (after all why mend something that isn't broken?) but I still maintain that I can go into most multi-cat households and find stress of one degree or another. I have tried to place some emphasis on preventative measures also, just to make sure your home remains harmonious.

I would strongly recommend that anyone facing a problem with their pet should read *Cat Detective* in its entirety. If you choose to dip in and look for a quick-fix solution you may end up referring to the wrong chapter. For example, don't think that reading Chapter 4 about house soiling will give you all the answers if your cat is urinating on the bed. You really need to read Chapter 7 on multi-cat households and inter-cat aggression, Chapter 5 on urine spraying and marking behaviour and also Chapter 9 on medical problems to give you a better insight into inappropriate urination. See what a complex animal the domestic cat is? Do not think for one moment that this self-help guide offers instant solutions!

Although *Cat Detective* is targeted specifically at those owners experiencing problems with their cats it doesn't mean that every cat person wouldn't benefit from some of the pearls of wisdom it contains. I am a staunch believer in prevention as well as cure and there is a great deal of advice in this book about limiting the chances of things going wrong in the first place. There are also answers to many frequently asked questions about general care, such as choosing a cattery, or car travel, that I hope all owners will find useful.

Throughout the book, for the purpose of clarity, I refer to cats as male, but do not feel this is discriminatory in any way; problem behaviour has no sexual bias.

In Chapter 4 onwards you will read detailed accounts of cases showing the cause of the problem and the programme that the owner followed to solve it. If you are currently experiencing a similar situation at home it may be tempting to adopt all or some of the suggestions to see if it helps your own dilemma. It is important to remember that the original programme was designed for *that* cat in *those* circumstances and it may not be quite right for you. However, I do try to describe it in as general a way as possible so that it can be adapted for other households. It's certainly worth a try.

As you read the book you will notice that many of the programmes have common elements, for example play therapy, and there is no question in my mind that all cats benefit from this sort of stimulation. However, the personality of your own cat will dictate the sort of games that you choose to play. Changing the way you behave towards your cat is also a common recommendation. You will soon see how often I ask owners to ignore their cat. If you see a case history that is similar to your own predicament then it may be worth trying some of the more general suggestions to see if they work for you. Don't lose heart if they don't because there may well be another element to your problem that is, as yet, undiscovered.

The important thing to remember is that, if your cat is exhibiting a behavioural problem, you are not fighting alone. There are good sources of advice and willing professionals to help you; you just need to know where to find them. However, before you seek the assistance of a third party, why don't you play cat detective? You may be surprised how much you can achieve yourself. Good luck.

CHAPTER 1

What Constitutes Problem Behaviour?

BEFORE WE EVEN START TO ADDRESS PROBLEM BEHAVIOUR IT is important to understand what we are up against. The dictionary definition of problem is 'anything that is difficult to deal with, solve or overcome'. Just remember that; we could be dealing with *anything*!

Normal behaviour performed in an inappropriate situation

Many of the problems that I am asked to address relate to normal feline behaviour. It is unreasonable to expect an animal not to behave naturally. Scratching, urine marking, climbing and even fighting, for example, all represent natural behaviour for the species. However, when a cat destroys an antique chaise longue by scratching or sprays urine into an electrical socket it is safe to say that this conflicts with normal domestic life in the average family home.

Inter-cat aggression is inappropriate if two cats start World War III in your front room but it certainly isn't abnormal or unnatural (read Chapter 7). Merely understanding this fact may make it easier to resolve the problem without apportioning blame to the aggressor. It is hard not to hate bullies since they would make such unpleasant human beings. However, they make extraordinarily effective cats when fighting to survive.

Predatory play is normal. Chasing a ping-pong ball or a feather on the end of a string is a pleasant activity that hones and refines hunting skills. When the target is the owner's hands and feet the behaviour is certainly inappropriate but it doesn't make the cat psychotic.

It is really important to identify those problems that arise from normal behaviour. In order to get a resolution it has to be remembered that the cat still needs to perform the behaviour, just somewhere more appropriate. Trying to stop scratching *anywhere* in the house, for example, is doomed to failure.

Examples of normal behaviour that may become undesirable are

- indoor urination
- indoor defecation
- urine spraying (marking territory with urine)
- middening (marking territory with faeces)
- inter-cat aggression within the household
- inter-cat aggression in the territory outside (either victim or perpetrator)

- some aggression to humans (for example, play 'aggression')
- scratching on furniture or carpets

Incompatibility of owner/cat personality

If some owners stopped to think about the problems they are experiencing it could be that their relationship with their pet falls into this category. Has your cat really turned out to be the right one for you or vice versa? Purchasing or acquiring a cat will always have an element of gamble associated with it. The actual decision-making can be quite unscientific; here are just a few examples:

- I've always wanted a silver tabby.
- I felt sorry for him.
- Nobody else would want a cat with a history of aggression.
- I was looking for a cat just like the one I had when I was a child.
- It was the only kitten they had left.

What a person really should be looking for is a cat with the right personality, background and age to suit the sort of environment and lifestyle that he or she can provide. All sorts of problems can occur if the need to match up compatible personalities is not taken into consideration. Here are some examples of relationships that probably won't work.

Unsocial cat/over-attached owner

This is often an acquisition based on colour, breed or any of the other coin-tossing methods of decision-making. The cat turns out to be a real loner and the new owner spends his or her entire time chasing it round the house to hold, stroke, squeeze and generally love it to pieces. This constant focus becomes extremely distressing for the cat and it develops all sorts of stress-related problems and weird and wonderful coping strategies to deal with it. The owner calls in a behaviourist to make the cat friendlier or stop it pulling all its fur out when what is really needed is a different cat (or a different owner).

Over-attached cat/over-attached owner

This may sound great on paper but it can be the ultimate dysfunctional relationship (see Chapter 10). Extremely compliant and caring owners focus totally on their cats. If they are matched with cats prone to insecurity then a strong dependency could soon develop. This is great until the owner in question actually wants a life outside the cat/human relationship. Poor pussy left all alone will potentially become highly distressed when separated and behave accordingly. Alternatively the cat may be a manipulative master of control and get the owner jumping through hoops just because it can.

Shy cat/noisy household

Cats from noisy and busy backgrounds are fine in noisy and busy houses. However, a shy and timid soul thrust into the average family home with 2.4 teenagers and a dog would probably dissolve in a heap or implode.

Over-attached cat/under-attached owner

I have experienced at least one case of this although, in my experience, it is relatively rare. An under-attached owner is possibly a good match for a potentially over-attached cat providing the cat has another focus in its life, such as going outside or activity indoors. However, the pairing is never perfect because, at the end of the day, you have a dissatisfied owner.

Unrealistic owner expectation

This is a very different problem from incompatibility because there probably isn't a cat in the land that would fulfil such an owner's expectations. There are a number of conditions that some people place on cat ownership and they usually have no bearing on a cat's natural needs, behaviour or desires whatsoever. For example:

- It is unrealistic to keep a cat exclusively indoors in a small bedsit, no matter how much a person needs company. There is nothing in it for the cat.
- It is unrealistic to expect a cat to sit quietly on a sofa twenty-four hours a day for the sole purpose of being there when the owner gets home. Some cats will do this but, not surprisingly, easily develop physical or emotional problems as a result.

- It is unrealistic to expect a cat not to hunt, fight or climb over the furniture. It's what they do.

Neurological abnormality

This represents that twilight zone between genuine behavioural problems and those with a neurological origin. The cat will undoubtedly exhibit pretty unusual and worrying behaviour but counselling or therapy alone would actually be pointless and inappropriate. These cases often call for a joint effort on the part of the referring veterinary surgeon and the behaviour counsellor. Sometimes it is useful to see a behaviourist to rule out a purely emotional cause before exploring further. I will discuss this in detail in Chapter 9 and show an example of a case that illustrates the point well. If you are fortunate enough to be referred to a behaviour counsellor who is also a neurologist then you stand a sporting chance of getting to the bottom of the problem. A history of trauma or disease is often relevant in these cases.

Neurophysiological problems relating to imbalances in brain chemistry can also be responsible for abnormal behaviour. The persistent repetition of sequences of movements that interfere with normal behavioural functioning is called stereotypy; a typical example of this would be pacing, weaving or over-grooming. Obsessive-compulsive disorders (OCDs), sometimes referred to as compulsive disorders, are also seen in cats and these usually need joint intervention from both behavioural and medical practitioners. I would also classify anxiety and depression as neurophysiological: potentially a very complex subject! Pica (the art of ingesting non-edible substances, as illustrated by the wool-eating Siamese) would also fall into this category. Cats that develop unusual seizure activity or a condition called hyperaesthesia – excessive physical sensitivity – should be purely in the domain of the veterinary neurologist, in my opinion. Certainly these are tough issues for owners to deal with on their own. I would strongly recommend that you don't attempt DIY in these cases, but I may be able to help you to identify such problems in Chapter 9.

Medical problems

Some problem behaviour results from medical conditions and the most obvious cause is pain. If your cat suddenly attacks you for no apparent reason, or his demeanour changes inexplicably, then the first investigation should be conducted by the vet. I have known many cats that have become bad-tempered or remote as a result of dental disease, for example. It hasn't affected their appetite but it has given them nagging unrelenting pain. It is amazing how a quick trip to the vet's surgery for the necessary extractions will result in a reformed character. There are a vast array of conditions that could potentially cause a change in behaviour, including

- urinary tract disease (inappropriate urination, urine spraying, over-grooming)
- bowel disease (inappropriate defecation)
- trauma/unspecified pain (aggression, inappropriate urination, increased sleep, behavioural changes, over-grooming, self-mutilation)
- hyperthyroidism (inappropriate defecation, night-time vocalization, aggression)
- arthritis (aggression, increased sleep)

- neuralgia (tail chasing, tail biting, self-mutilation)
- flea-allergic dermatitis (over-grooming, self-mutilation)
- other allergies, e.g. food (over-grooming)

CHAPTER 2

What to Do When Things Go Wrong

THIS BOOK IS NOT INTENDED TO BE AN ALTERNATIVE TO professional help; hopefully its role is to show you what to do when things go wrong. There is a definite sequence to follow when your cat starts to exhibit worrying and inappropriate behaviour. I am always emphasizing the role of the veterinary surgeon in dealing with these issues; the surgery should always be the first port of call. Many problems have a physical origin and a visit to the vet for the necessary treatment will often provide a resolution. However, if the vet gives your cat a clean bill of health it would be entirely appropriate to look for a psychological origin to the behaviour. The next few chapters will provide the information you need to make an informed judgement about possible causes; the solution often comes easily once you know why your cat is unhappy.

- Remember, your cat is not naughty. He is unhappy because something has gone wrong with his world and he's dealing with it the best way he knows how.
- Some problems resolve quickly without intervention. Most, however, come and go with varying intensity, so it is easy to let things drag on without facing the fact that a problem exists. If your cat behaves inappropriately, don't wait before doing something about it. You are merely making a solution more difficult to achieve.
- Visit the veterinary practice to see whether there is a medical cause or component to the problem, and then, if medical problems have been ruled out, examine the history of the behaviour using the questionnaire in Chapter 3.
- Make the necessary adjustments to the way you treat the cat to help to change the behaviour.
- Make the necessary adjustments to the environment to help to change the behaviour.
- Monitor the progress by keeping a diary to see if the behaviour improves or deteriorates. It's amazing how easy it is to forget how bad the problem was last week, let alone last month!
- If the problem has not resolved or improved significantly within the first month then speak to your vet about a referral to a behaviour specialist. Remember to ask your vet for medical histories on your cat/s and a written referral, as the behaviour counsellor will undoubtedly need these.

Employing a counsellor

All reputable pet behaviour counsellors work on referral from veterinary surgeons only. There are various associations and organizations relating to the world of pet behaviour counselling and they all have websites, so you can do your own research, but although experience and qualifications are important the best advice would be to have a personal recommendation from your vet. Even if a behaviour specialist works predominantly with dogs it doesn't necessarily mean he or she will not be helpful to you. I know many multi-species practitioners and they are extremely effective. Here are some important criteria for you to check before you make the final decision to employ a behaviourist.

- Do they make house visits? This is essential in my opinion for dealing with cat problems as the environment and household dynamics are so influential.
- Do they have professional indemnity insurance? It is always advisable to establish this from anybody giving paid advice relating to your pet's well-being.
- Do they confirm all instructions in writing? There can be a mind-boggling array of information imparted during a consultation so it's important to have a written report afterwards containing all the relevant recommendations.
- What fee do they charge? This can vary enormously depending on a number of criteria so get a quote before you make a decision. Don't necessarily go with the cheapest or the most expensive; stick with the one who has been recommended and whom you feel confident about. Many pet insurance companies cover behaviour referrals so it's worth checking; remember that before they accept your claim they

will have very specific conditions about the referral process from your vet.

- Do they provide continued support after the consultation? This is essential as things can change and it is possible the programme of therapy needs to change too. It's also good to know there is someone on the end of a telephone who understands the problem.

Preparing for a visit from a cat behaviour counsellor

Once your vet has referred you and your cat to your chosen therapist it is important that you get the maximum benefit from their initial visit. We all feel we can remember everything there is to know about our cats, but you may be surprised when the big day comes how vague you will be. You need to concentrate and give as accurate a history as possible of the problem and your cat's background. The quality of your information will strongly influence how effective your behaviour counsellor will be with a 'diagnosis' and possible solution. Here are a few tips to help you to get the most out of the visit.

- Ensure you have all the relevant paperwork from your vet that the behaviour counsellor requested. This may already have been forwarded direct prior to his or her visit.
- Collect together all the paperwork associated with your cat's vet bills or original adoption or purchase. This will help you get the order of events right and prompt your memory about dates.
- Make notes before the visit about important dates and facts and maybe scan through all the questions in Chapter 3 to

make sure you have a story that both you and your partner can agree on. Arguing and contradicting each other about your cat's history is time-consuming and can be embarrassing for everyone, even the cat.

- If you have young children, it may be useful for a relative or friend to take charge for a few hours. Stopping to entertain the child or change nappies can be distracting.
- Try to keep your cat (or cats) indoors for the duration of the consultation. It is important for the counsellor to observe the interaction between them. He or she will also want to see how you interact with them.
- Have a pen and paper ready in case you wish to make notes. There may be some things that you can implement straight away without waiting for the behaviour counsellor's written report. If you are already emotionally and intellectually drained by this point, just sit back and listen; everything will be confirmed in writing later.
- Finally, never be nervous when a cat behaviour counsellor comes to your home. He or she is not there to point the finger of blame but to help. Enjoy the experience of learning more about what makes your cat tick.

Role of a cat behaviour counsellor

The cat behaviour counsellor will probably need to spend a couple of hours with you at home. During that time you will be asked numerous questions about your cat's background and lifestyle. The history of the problem behaviour will also be discussed in some depth and the counsellor will take the opportunity to observe the cat. Everything will go very smoothly because you are suitably prepared having followed the suggestions above. Towards the end of the discussion you will probably be given an explanation of why your cat is exhibiting the problem behaviour. Once the reason has been established the plan of action will be formulated to resolve the problem. The behaviour modification (or behaviour therapy) programme will vary depending on the nature of the problem and on the practical considerations of your household, but it usually consists of a number of changes to the environment that will assist in modifying your cat's behaviour. For example, if your cat is urinating inappropriately indoors you may need to provide a particular type of litter tray in a specific location. You may also be asked to follow a different cleaning regime and to stimulate your cat more with games and toys. There may be a need for you to react differently to your cat. You may be advised to pay it more or less attention or to show different body language in response to certain behaviour.

Once the plan has been created, it is important that you discuss with the counsellor any areas where you feel there is confusion. Some concepts are quite tricky to grasp so don't worry if you don't instantly understand everything that is asked of you. You may also feel that you are unwilling or unable to fulfil certain requirements. I would rather a client told me at the time than failed to do something and ended up with no resolution to the problem. Behaviour programmes are all about compromise so don't feel guilty if you have to negotiate a little at the beginning. It may represent the difference between success and failure.

The counsellor will confirm everything that has been discussed during the consultation in writing and your vet will also be kept informed of the proposed plan of action. The counsellor will recommend a period of time during which you should follow the programme and report your progress on a regular basis. Some counsellors will contact you at a pre-arranged stage and others will rely on you to contact them. Either way it's important to choose a counsellor who will provide continued support for at least six weeks after the initial consultation.

Combination therapies

A number of other things may be recommended to work alongside the behaviour therapy programme. Herbal remedies, supplements and homoeopathic treatments might be suggested, but it is essential to check with the referring veterinary surgeon before using them. Your counsellor, unless a veterinary surgeon, is not technically qualified to prescribe complementary medicine. I have worked with homoeopathic vets in the past and I tend to favour the use of flower essences and synthetic feline facial pheromones in many of my programmes. I haven't had great success with either in isolation but they seem to work well in conjunction with the behaviour therapy. It is interesting to note that many owners attribute the success of the treatment to the flower essences or pheromones rather than believe that interactive and environmental changes can have such an amazing effect. Since I tend to apply a rather 'belt and braces' technique to my programmes I'm just glad they work.

Homoeopathy

Many modern veterinary practices are embracing homoeo-pathy as part of their service. This branch of medicine works on the concept that any substance that causes symptoms in humans (or animals) can be used in the treatment of illnesses that show the same symptoms. The amount of the remedy used is incredibly small and administered in tablet, powder or solution. Homoeopathic remedies can be useful in the treatment of urine spraying, stress-related cystitis, over-grooming and many other distressing problems in conjunction with behaviour therapy.

Flower essences

The principle behind healing with flower essences is similar to homoeopathy. The theory is that each single-flower solution has an 'imprint' that is the vibrational character of that flower. When it is ingested it causes cells in the body to vibrate at the same level and alter the balance of energy. It is a complicated concept to embrace and the efficacy of flower essences has yet to be proved scientifically. However, they have been around for years, with modern systems (the earliest developed by a British physician called Edward Bach) designed for use in humans. I have been using them to help animals since 1988. When I worked with the RSPCA we often gave a drop of Bach's Rescue Remedy to injured wildlife prior to treatment. I was always surprised at the difference in recovery between patients that had received the drop and those that hadn't. I must admit that the essences' mode of action is a bit of a mystery to me but there is plenty of anecdotal evidence that they can really make a difference.

There is a wide variety of flower remedies and each targets a particular negative emotion that your pet may be experiencing. The commercially produced essences, as I have mentioned, are intended for treating humans, but there are several very good books available that give details of dilutions and treatment suggestions for a range of domestic animals. The flower essences are preserved in a grape alcohol and if your cat is taking certain medication, such as metronidazole, it may induce vomiting. Otherwise I have found them to be perfectly safe – but always remember that their use should be approved or prescribed by your vet.

Feline facial pheromones

A pheromone is a chemical substance that an animal produces to signal a specific message to other members of the same species. All cats have glands in their cheeks that secrete a unique pheromone. They use this scent to mark their territory and the smell gives them a sense of security and reassurance; it's a little like aromatherapy for cats. These pheromones have proved extremely useful in behaviour therapy as they can provide a positive scent message at times when cats are feeling anxious. Research indicates that cats are reluctant to scratch or spray urine in locations where these facial pheromones are present. I use them in combination with behaviour therapy in a number of different situations, such as

- urine spraying indoors
- excessive scratching indoors
- some instances of inter-cat aggression
- car travel
- cattery visits
- moving to a new house
- introducing new furniture into the home
- anxiety-related inappropriate urination
- general stress

Collecting a cat's facial pheromones

An individual cat's scent can be collected and used where necessary. This can be done in four easy steps.

1 Use a small soft natural fibre cloth or fine cotton gloves (available from pharmacies).
2 Stroke firmly around the cat's head, using the cloth or gloves with particular emphasis on the cheeks, chin and forehead.
3 Rub the impregnated cloth/gloves against the surface that has been scratched.
4 Repeat the process until the cat's scratching is redirected elsewhere.

Synthetic feline facial pheromones

A part of these facial pheromones is common to all cats and a synthetic version is available in spray and diffuser form (the latter plugs into an electrical socket) from veterinary practices.

'Kitty Prozac' and other drug therapy

I must admit to being strongly opposed to drugging cats that exhibit problem behaviour although I do accept that, in extreme circumstances, there is little alternative. Genuine neurological problems can be greatly assisted by short- or long-term medication. However, I believe most behaviour problems can be resolved without resorting to potent and toxic medicines. The reason why they are used relatively frequently these days is that the alternative is often to remove the cat and re-home it elsewhere and the owner cannot face that decision. This is a complex issue and I would always recommend discussing the matter thoroughly with your vet before embarking on any drug therapy for problem behaviour.

Tips about drug therapy

- Always consider drug therapy as a last resort after exploring all other avenues.
- Medication can only be prescribed by a veterinary surgeon.
- Many of the drugs used to treat anxiety or any other emotional state are not licensed for use in cats. Your vet may ask you to sign a disclaimer.
- These drugs often need to be administered for a relatively long period to have the desired effect.
- Don't consider drug therapy on its own to treat a behavioural problem. All drug therapy should work alongside a programme of behaviour therapy that has been devised by a pet behaviour counsellor or a veterinary surgeon with a particular interest in behaviour.
- Blood tests should always be taken prior to and, in some cases, during drug treatment to monitor liver function.
- Follow advice concerning dose rates carefully. Never stop drug therapy abruptly without consulting your veterinary surgeon.
- Keep these drugs away from children and vulnerable adults.

All of these considerations may go some way to ensuring our cats do not become as reliant on 'happy drugs' as the human population.

I tend to limit my own endeavours to those described above but there are always other possibilities for clients to explore if they wish to investigate every possible treatment in combination with behaviour therapy. Many modern veterinary practices are fairly progressive and show great enthusiasm for and knowledge of alternative and complementary therapies; it is very encouraging to see many practices now offering physiotherapy, hydrotherapy, homoeopathy and acupuncture. Where behavioural problems are caused by underlying medical conditions they may well be alleviated by a combination of conventional and alternative therapies, but remember that any complementary treatment should be carried out on the recommendation (or with the blessing) of your veterinary surgeon.

Acupuncture

Acupuncture has been used to treat animals as well as humans for four thousand years but it has only been actively pursued in Western veterinary medicine over the past thirty years. The interpretation of exactly how acupuncture works varies somewhat between the oriental perspective and that of the Western world but both apply pressure (using needles or other techniques) to the same points around the body to relieve pain, stimulate healing or generally improve well-being.

There are some inappropriate behaviours that could be assisted using acupuncture, including
- chronic idiopathic diarrhoea (resulting in inappropriate defecation)
- cystitis (resulting in inappropriate urination)
- neurological disorders (manifesting themselves in seizures or self-mutilation)
- dermatological disorders (resulting in over-grooming or self-mutilation)

- painful musculoskeletal conditions (causing aggressive behaviour, over-grooming, self-mutilation)

There are some drugs, including steroids, tranquillizers and anticonvulsants, that are contraindicated with acupuncture so it is important that the practitioner gives your cat a thorough medical examination and takes a full history prior to treatment.

Reiki

Reiki is a 'laying-on of hands' healing treatment and many owners feel that their pets have had genuine relief from physical and emotional problems using this form of therapy. Whilst I have no personal experience of the benefits of Reiki I am keen to embrace anything that is non-invasive and appears to provide relief and solace to the cat and owner.

Tellington Touch

Some cats can be very resistant to handling and we can be rather clumsy in our approach sometimes. A method of handling devised by a Canadian called Linda Tellington-Jones teaches owners to touch their cats in a specific way that promotes relaxation and improved bonding between owner and pet. Tellington Touch also utilizes soft artists' brushes and feathers to emphasize the importance of lightness of stroke. Even cats that are normally reluctant to be touched seem to find this method of approach acceptable.

Clicker training

When you first consider the concept of training a cat it does sound rather as if you are attempting the impossible. I have even been known to make a rather feeble joke about the manual of clicker training for cats being a very short book, merely containing the words:

Step 1   Don't!

Step 2   Get a dog.

However, when you look seriously at clicker training you will see how useful the process can be. The method uses a device called a clicker: a small plastic and metal object that produces a novel sound. It is used together with a treat or 'positive reinforcer' to reward and teach new behaviour. Potentially, it can stimulate a bored house cat, prevent aggression, extinguish undesirable behaviour and even improve the relationship between cat and owner. In order to get the best results from this method you really need a hungry cat. The positive reinforcer to teach new behaviour is usually food and many cats have a permanently satiated appetite. If you really want this to work for you and your cat, it is best if you reduce the amount of food you are providing throughout the day (it may well be too much anyway) and find a special treat that your cat really loves.

CHAPTER 3

Playing Cat Detective

BEFORE EMBARKING ON A VOYAGE OF DISCOVERY ABOUT WHAT makes your cat tick and why he is currently behaving badly it's worth finding out how a cat acquires its unique personality. Whatever is happening to your cat is not just circumstantial. It is really about your cat's own personal reaction to those circumstances, and his ability to cope under pressure. Change the environment and things would be very different; change the cat and the problem behaviour wouldn't arise in the first place.

The development of a cat's personality

Anyone involved with cats will know that every individual has

a unique character. Adult cats and kittens vary enormously in their friendliness towards humans; even kittens from the same litter can differ considerably. If you watch them, it will soon be apparent that one is behind the sofa, one is climbing up your trouser leg, one has its head firmly embedded in a tissue box and two are chasing each other up the curtains. They are all experiencing the same environment and the same social contact but each one's response is influenced by its own unique personality.

The importance of early socialization

A cat's personality is formed from both genetic and environmental elements. Genes 'programme' an individual with the potential to react in a certain way in certain circumstances. The individual's life experiences then determine whether that behaviour is ever actually expressed and to what level. The most significant behavioural and emotional development takes place between the ages of two and seven weeks. During this time, referred to as the sensitive period, positive exposure to humans and other species develops the kittens' potential to form social bonds with them. This process is referred to as 'early socialization'.

Research conducted into the sensitive period concludes that handling by a number of different people during this time will tend to increase the kittens' sociability towards humans. Positive exposure to the environmental challenges of modern twenty-first-century living, such as children, dogs, noisy household appliances, different locations and even car journeys, will better equip the individual to cope with life in the future.

Whilst the period between two and seven weeks is usually outside the new owner's influence it is still important to stimulate older kittens as much as possible with a variety of challenges. Lessons and positive associations can be learned at any age providing the appropriate genetic 'blueprint' is present.

Social play

Interactive play with siblings has a role in the development of later social skills. Solitary kittens do eventually form attachments but are generally slower to bond than normally reared kittens. Neither do they learn to hold back when they play fight if they target human hands rather than siblings. A person cannot possibly teach the boundaries of acceptable levels of physical force as well as another kitten. Uninhibited biting and grabbing can easily lead to an individual's being labelled 'aggressive' when all he is doing is being playfully boisterous in the only way he knows how (see Chapter 6).

Categorizing personality

There are numerous ways to scientifically categorize character and personality, but there are two basic models, 'excitable and reactive' and 'slow and quiet'. Variations in excitability and timidity may well be caused by inherited differences, such as the amount of adrenalin released when faced with a challenge. (Cats, like humans, have an instinctive 'fight or flight' response to danger, fuelled by the release of adrenalin to pump blood into the muscles and away from the non-urgent organs such as the gut.)

Certain breeds are often described by their temperament. For example, Siamese are considered to be sociable, affectionate, sensitive and vocal. They are also prone to eat wool, spray urine on their owners to get their attention, and pull their fur out – but that's only the minority! Burmese are traditionally assertive and outgoing with a tendency, every now and again, to be aggressive and territorial; and the Persian is placid, although occasionally showing aggression during grooming sessions or appearing to forget where you put the litter tray. The consistency of these descriptions must imply that the characteristics are inherited.

Are there stereotypes in the cat world?

I truly believe that every cat is an individual but, just for fun, I tend to think that there are stereotypes in the cat world. It may be useful to look through the list below to see which one best describes your pet. The solution to most problems has to take into account the fundamental character of the cat in question, and this is where the whole concept of behaviour therapy becomes complicated. For example, the programme required for a highly attached cat would be very different from one designed to treat a cat that does its own thing and interacts with the owner only at mealtimes. Some cats, too, are influenced significantly by their owner's mood and behaviour, whilst others appear oblivious. However, it is true to say that certain types of cats are more prone than others to certain types of problems, and this may be a very helpful piece of information when you are trying to unravel your own problem puss's mind. Although I refer to all these cats as male they are just as likely to be females; there don't appear to be any stereotypes that are gender specific (although a greater proportion of Cuddle bunnies seem to be male).

The Lodger

This cat does his own thing, comes and goes, never complains, but is almost always indoors and fast asleep at night. The Lodger is the cat for everyone who ever said 'cats are less of a tie than a dog'. The Lodger loves you best when you are opening a can of cat food or making tuna sandwiches. He will sleep on your lap at night if you are warm and promise to remain still.

Inscrutable

Whatever this cat feels about life you would never know. Inscrutable's expression never changes; he always appears content but there is plenty going on in his head – watch for subtle changes in his behaviour. Inscrutable is friendly enough but also happy to disappear and patrol his territory.

Beanbag

This cat gives the appearance of being stuffed with beans. He sleeps on his back with his legs in the air and the explosion of a major incendiary device nearby will merely result in a flick of his whisker. Beanbag tolerates all newcomers into the home with a huff of resignation and doesn't really mix with other cats unless absolutely necessary (e.g. mealtimes etc.).

Twitchy

Twitchy jumps at the merest sound or movement but appears perfectly relaxed otherwise. He sleeps a lot indoors and tends to go out mainly when you are gardening or chatting with the neighbours. He will sit beside you but would rather not sit on your lap.

The Comic

Who said cats don't have a sense of humour? Living with a Comic provides an endless source of amusement, what with his apparent tendency to sleep in absurd postures and his propensity to fall off things with monotonous regularity. This cat will play with anything and seems to have boundless energy as he skids across the kitchen floor in pursuit of something ghastly he's just found under the fridge. Owners often describe the Comic as 'a sandwich short of a picnic' but he's merely enjoying life.

Jekyll and Hyde

This is the most loving and affectionate cat to his owner but a thug and a bully with other cats. He appears to take sadistic pleasure in fighting, torturing and terrorizing other cats. Picking on the elderly is particularly entertaining. He will break into other people's homes to steal food and intimidate the resident cats. These cats are best not disturbed by protective owners because they don't discriminate between cats and humans when they are on a mission of destruction and mayhem. Dare I say that many Jekyll and Hyde cats I have known over the years have been Burmese?

Fickle

This cat demands affection on occasion but is distant most of the time. He is happy to have attention but very much on his own terms. Sometimes he ignores you and then he can't get enough of you; very confusing for those owners who like to know where they stand in a relationship.

Schizo

This cat rubs round your legs and then has your face off if you dare to touch him. Schizo is the master of mixed messages; he will often sit on your lap and then bite and scratch you if you stroke him for a second too long. Owners believe that these cats must have been ill-treated as kittens but this is rarely the case. They are merely objecting to the fact that their owners are such rubbish at speaking cat.

Scaredy-cat

Scaredy-cat might as well be a dust mite, considering the time he spends under your bed. He tends to come out at night or when you are sitting down, and heaven help you if you move. He will do the four-minute mile upstairs when the doorbell rings, not to be seen again until the following day.

Cuddle bunny

This cat seems to have Velcro on his tummy because he spends most of his time stuck to the front of your sweater. He follows you everywhere, sleeps in the bed with you, and dribbles, purrs and treads on your stomach constantly. He is definitely mummy's boy, prone to losing his appetite when you are away (but then of course you don't go away, do you?). He would undoubtedly pine to death in a cattery, or so you believe.

Snooty

This cat looks as if he cannot believe your audacity when you dare to touch him. The whole experience makes him feel physically sick. Snooty disappears to a private place with monotonous regularity because he's trying to get away from you. However, please don't forget to give him a bowl of biscuits before you leave the room.

Struggler

This cat is really cute and sweet until you try to give him a pill or an injection. He will then develop four times the normal quota of legs, teeth and claws and contort his body to ensure maximum penetration of any human flesh in the vicinity. Whatever anyone is trying to do to him they will never achieve it unless he is anaesthetized or dead. Trust me, this cat has 'CARE' written all over his medical records at the vet's. Many of the other stereotypes listed are part-time Strugglers.

The Ringmaster

He will get you jumping through hoops just because he can. This is the cat that demands your total compliance and has an armoury of devious ways to ensure you obey at all times. You love him dearly, totally unaware that he is probably laughing behind your back.

The Old Soul

Please excuse me for sounding fanciful but every now and then our lives are touched by Old Souls and it would be a shame not to include a rather special stereotype in this list. These are cats that, no matter what you throw at them, seem to possess a wisdom and presence more at home in a Tibetan monk than a five-year-old moggy. Old Souls tend to be legendary creatures that turn up on your doorstep and adjust completely and effortlessly into your life. No one knows where they come from but they are always remembered long after they are gone. They will go for walks with you, meet the children from the school gates or raise your spirits when you are down. My own dear Hoppy was definitely an Old Soul. The courageous black and white cat Simon from HMS *Amethyst* was probably one too. These cats have definitely been here before.

All of these stereotypes have genetic and environmental elements to their character that make them what they are. For example, Snooty, Struggler, Fickle and Schizo may all have had poor early socialization so their ability to interact with humans is lacking to some degree. Scaredy-cat was probably born that way and any attempts at socializing would fall on deaf ears. Cuddle bunny and the Ringmaster have both had good experiences of humans but choose to approach interaction with them differently. Cuddle bunny develops learned helplessness and becomes completely unable to cope without his owner, and the Ringmaster discovers how easy it is to control a human being from the comfort of a radiator hammock. Never worry about an Old Soul; they don't get behaviour problems!

✾   ✾   ✾

You have already started the long and tortuous journey towards objectively assessing your cat's personality. Now we need to look at all the questions you should ask yourself to get to the bottom of the problem that you are currently facing. Whenever a cat behaviour counsellor works with a patient he or she will always ask exhaustive questions about the cat's history and background. They will often do this before they even start to discuss the problem behaviour. The answers give everyone a better understanding of why your cat has reacted in a particular way to its current dilemma. Listed below are a number of examples of the type of question that will be asked and a brief description of why the answer may be relevant. If you ask yourself all these questions (get the whole family to contribute to the answers) it will probably help you to understand the problem and hopefully identify possible causes.

It is probably too complex to try to look at every possible behavioural problem you may experience and list relevant questions accordingly. I have therefore concentrated on house soiling, urine spraying, aggression towards other cats and aggression towards humans. Many of the questions listed are certainly worth asking no matter what problem you are experiencing, since identifying what is wrong with your cat's world is always going to be important.

Behaviour questionnaire

The first section is going to be relevant to every problem and provides a good background to your cat's particular circumstances. It will give you a great opportunity to view the situation rather more objectively than normal.

How many cats in the household? If you have a multi-cat household (two or more) then it may be worth looking at the relationship between the cats to see if this could be a cause of the problem. (See Chapter 7 for further details.)

What are their names, ages, sex and personalities (bold, sociable, timid, independent, playful, etc.)? You know your own cats so this seems daft but it is a good way to actually think about how you would describe their personalities. Use the stereotypes listed earlier if it helps you focus. This may not hold the key to the problem behaviour but understanding the character of the individual will certainly help with the solution. For example, a Cuddle bunny may benefit from being encouraged to be more self-reliant and a Schizo may feel more relaxed if you learned to speak cat a little more fluently. It may be interesting also if you find the cat was between eighteen months and four years of age when the problem started. This is when cats mature socially, often heralding the start of issues with other adult cats. (See Chapter 7.)

Do you have other pets? If you have other pets it is worth checking to see if their arrival coincided with the start of the problem. Dogs can be a problem but cats soon learn they have the upper paw. In my experience, the introduction of another cat is more problematical. If you have other cats it's important to look at the relationship between them. At this stage it is always going to be a subjective assessment but how do you perceive their relationship? Are they tolerant or really sociable? Is one threatening and intimidating or do they lead separate lives?

How many people are there in the household and have the dynamics changed recently? Cats can be sensitive to changes within the human household. Has a member of the family left home or has a new person been introduced?

Who interacts with the cat/s the most? This person will potentially have the greatest impact on the cat's behaviour, and the influence can be good or bad. Not all interaction is actively encouraged by some cats and being paid too much attention can be distressing.

Does the cat have access to outdoors? Cats that are confined permanently indoors are particularly sensitive to everything that happens within those four walls. Those that have access outside may be intimidated by other cats in the territory. Is there a new cat on the block? Indoor cats need lots of extra stimulation to compensate for their confinement. Don't forget that the devil makes work for idle paws.

How does your cat gain access outdoors (cat flap – manual or magnetic exclusive-entry system)? Whilst cat flaps look like a convenient addition to the home they actually represent a vulnerable point in its defences. Don't forget that a cat's home is its castle and a cat flap is potentially a dropped drawbridge. Your cats may be able to come and go but so can any other in the neighbourhood. If you are a 'doorman' for your cat and allow access on demand through a door or window you will create the perception of a more secure environment.

Is there usually someone at home during the day? Cats don't necessarily need us with them twenty-four hours a day, but if there is someone at home cats can become more dependent on the company.

What is your cat fed and how often? Just like us, cats *are* what they eat. Is your cat eating excessively or has his appetite decreased? Is he fed on demand? If so, could this indicate that you are a little too much at his beck and call?

Where does your cat eat? Many cats eat in the kitchen or surrounding area for obvious reasons. Is that bowl in the corner the most exciting way to acquire food? Cats are designed to hunt so where's the challenge in two meals a day in a predictable place? Food for thought. If you have more than one cat are you expecting them to share a bowl? Do you notice friction between them at mealtimes?

Where does your cat drink? Your cat may not be drinking enough water if there is merely a bowl beside the food. When cats are hungry they hunt for food. When they are thirsty they look for water; the two are rarely found in the same place yet we put them together for our cats. Does your cat drink outside, or from the tap, the lavatory or the glass on your bedside table, by any chance?

Is your cat suffering from any illness, disability, or disease? Many medical conditions can influence behaviour. Make sure your vet has checked him over to monitor any deterioration or complicating factors.

When did the problem start? This may surprise you when you really look back. The longer you have been tolerating the problem the more difficult it will be to find a solution. However, it still isn't impossible, so don't despair.

Did it coincide with any environmental disturbance (building work, change of litter etc.)? Remember that everything that happens within the home has a potential impact on your cat's behaviour and emotional well-being. A new floor or carpet may be a blessing for you but it may represent a major trauma for your cat, particularly if you have replaced your carpet with wood or laminate flooring. If you have changed your garden recently to a low-maintenance concrete and pot paradise you may have inadvertently destroyed your cat's preferred toilet area.

Are there any other changes in the cat's behaviour since the problem started? Cats are creatures of routine so never underestimate subtle changes in your cat's habits. Has he taken to sleeping under your bed rather than on top of it? Is he sleeping more or less? Is he eating and drinking more or less? Is he grooming more or less? Is he rubbing his face on furniture more than usual, or scratching indoors? Is your cat more attentive or more remote with you since the problem started? Is he spending more or less time outdoors? All of these changes could indicate that he is distressed by the presence of another cat and he is feeling insecure, even in his own home. They could also indicate that your cat is ill so a check by the vet is always advisable.

How frequently is the problem occurring? Some house soiling for example will occur in cycles, particularly if the underlying cause is a chronic cystitis. There may also be a pattern to aggressive behaviour, urine spraying or inter-cat aggression. Is it worse in the spring, for example?

Has the frequency increased with time? House soiling will often escalate and you may be able to see a pattern emerge if you look at what has been happening within the home. Have you become increasingly intolerant or distressed yourself by the problem? Often these situations can be horribly self-perpetuating as both owner and pet become increasingly anxious. Many other behavioural problems will increase in intensity or frequency as the poor cat gets locked into the stress of his circumstances and seems overwhelmed. Other problem behaviour that is more undesirable than unpleasant for the cat, such as certain types of aggressive behaviour, can increase because of the obvious rewards the cat experiences when he does it.

Have you visited your vet for advice regarding this problem and to check for any medical problems? It cannot be emphasized enough that this should be your first course of action. You can't separate your cat's physical and emotional well-being and many conditions or diseases will undoubtedly affect his behaviour. If you have been given specific behavioural advice by your vet it is important to make a note of what you have done and when just in case you need professional help in the future.

Specific questions regarding house soiling

Are there litter trays available? Even if your cat has access to outdoors there may be occasions when he is intimidated outside. The last thing he will want to do is squat in the flower bed if he thinks the tomcat next door will jump on him. If your cat lives exclusively indoors the provision of trays is essential but you still need to take care that the facilities you provide are right for your cat.

If so, how many, what type and where are they located? Have you moved the tray or replaced it with a new one? If you have a problem you will need one tray per cat plus a further one, in different locations (see Chapter 4).

What type of litter material do you use? Some cats are extremely fastidious and won't tolerate hard wood pellets or silica beads, for example. Always try a sand-like fine grain substrate (there are plenty on the market). All domestic cats are descendants of the African Wild Cat and they gravitate towards sand (if you have ever had a pile of builder's sand on your driveway you will know how true this is).

What is the cleaning regime for your litter trays? If you have a covered litter tray it is easy to think 'out of sight, out of mind' when it comes to cleaning. Do you clean your trays daily? Are you masking the smell with strong litter deodorant? Would you use the tray if you were a cat?

Does the cat spend a lot of time in the tray (scraping and digging)? Those cats that 'pee and go' or balance precariously on the edge of the tray may have an issue with the type of litter you are using. If they spend hours scraping and digging they may just be illustrating their fastidious nature or trying to prove a point to the other cat/s in the house.

Does your cat cry prior to using the litter? Your cat is nervous and reluctant to use the litter tray for some reason and would really rather not go there at all. Is he in pain when he goes?

How often is the litter tray used daily (urine and faeces)? If your cat uses the tray frequently for small amounts of urine there could be a problem with his urinary tract. If he goes once every two days he is definitely retaining urine. This could be a sign of stress or reluctance to use the tray for some reason.

Where is the cat soiling? If your cat is peeing indiscriminately throughout your home and not returning to a previously used area then I would be highly suspicious of a medical problem. Even cats who soil in the home tend to favour the same two or three areas every time (until things get really bad). It is worth looking at the sort of location where your cat is soiling. For example, is it a corner somewhere quiet, or do all the chosen spots have the same sort of surface underfoot? Keeping an open mind at this stage is essential. Your cat may be marking territory with urine or faeces rather than merely relieving himself. (See Chapters 4 and 5.)

Is your cat urinating, defecating or both? It is more common for cats to urinate inappropriately than to defecate as a result of stress. If they are peeing and pooing then the problem is likely to be a litter tray aversion or a reluctance to go outside because of a local cat bully.

Does the cat still continue to use the tray intermittently? It is possible to have an intermittent aversion to a litter facility. Your cat could use it only when it is freshly cleaned, implying that he is rather fastidious and would prefer it to be cleaned more frequently. He may even stop using it when your other cat has made a deposit. Covered trays can also hold in unpleasant gases from urine and they really should be cleaned out regularly despite the fact that *you* can't smell it.

Does the cat continue to eliminate outdoors intermittently? If there is a bully cat out there he won't always be present. Your cat will know safe and dangerous times and may be eliminating accordingly.

Has your cat been involved in any disputes outdoors with neighbouring cats? This would confirm your suspicion that there is friction between your cat and another outside. It may also account for a reluctance to go outdoors as much as before.

Does your cat have a history of cystitis or urinary tract problems? There can be purely physiological reasons why cats develop cystitis but many cases are stress-related (see Chapter 4). Any pain or discomfort associated with urination can lead to soiling. Cats might become averse to using trays because of the pain experienced whilst there, or just need to pee too frequently to care whether they use the available facilities or not. A check at the vet's surgery will rule out any bladder problems.

How are you cleaning the soiled areas? Cats will often return to an area due to a residual smell left from previous urinations in the same spot. Many owners immediately reach for bleach-based cleaning products to guard against infection but this can be counter-productive. Bleach will break down to ammonia compounds just like urine and might even reinforce the smell you are trying to eradicate. Many cleaning methods just aren't good enough to remove persistent and invasive urine from carpets. A carpet acts like a wick, drawing the urine down through the underlay to the concrete or floorboards beneath. (See Chapter 4 for more information on cleaning.)

Are you punishing the cat? Well-meaning fountains of feline knowledge are still advising owners to punish their cats when they are found soiling. Better still, rub their noses in the relevant deposit! Such actions are totally pointless and will merely further damage the relationship between cat and owner and increase your cat's underlying anxieties. Whilst I fully understand the frustration engendered by constant house soiling I would recommend a visit to the gym instead to 'vent your spleen'.

Are you using deterrents? Various household items and foods are recommended as suitable deterrents to prevent your cat from soiling. In many houses I have had to negotiate assault courses of tin foil, pine cones, pepper, orange peel and cat biscuits stuck to cardboard. If deterrents are effective at all (and frankly most of them are not) they will merely send your cat off to soil in a new and previously undamaged location. There is no easy fix to these problems; you have to identify the root cause.

Specific questions regarding urine spraying

Are you aware of any new cats in the area (new neighbours etc.)? It's worth asking around; just because you haven't seen any other cats doesn't mean they are not there and causing problems.

Has a strange cat entered your house through the cat flap? This is the cat equivalent of aggravated burglary. Your cat may have appeared fairly laid back when it happened but that was probably an act to avoid being beaten up.

Does your cat spray urine indoors? This indicates that your cat feels insecure or anxious about something within the environment. It is usually a 'cat thing' and others in the house or outside are responsible for his anxiety. Some cats spray urine for rather more complicated reasons (see Chapter 5).

If your cat does spray urine are the amounts small (i.e. 2–3 mls) or large? Cats would normally pass small amounts of urine for marking purposes to enable frequent marks to be made. If your cat is emptying his bladder at the same time it may indicate he is retaining urine, highly stressed or suffering from urinary tract problems.

Does your cat scratch indoors, and if so where? This can be a marking gesture as well as general claw maintenance (see Chapter 5). If there is excessive scratching in one or two specific areas it could indicate that your cat is under pressure from other cats, either within the household or invading from the outside.

Specific questions regarding inter-cat aggression

Is your cat displaying aggression to other cats? If so, you would be well advised to ask all the other questions in the two previous sections because friction between cats rarely occurs in isolation. You will probably be experiencing some soiling or urine spraying or even be struggling with obese cats that pull their fur out (see Chapter 7 if you are now confused).

Is the aggression towards one cat in particular? Aggression is often targeted only towards those who are easily intimidated, either inside the home or in the territory outside. Even if your cat hates all others you will probably still find that he will pick on the one that is the most reactive to his aggression.

Is the aggression active (fighting) or passive (postural)? Don't forget that aggression can often take the form of subtle intimidation and psychological threat. If you constantly see your cat through rose-coloured spectacles then you just won't see him torturing his brother. If it makes it easier, don't watch the alleged aggressor but look at the reaction of the victim in his presence. This may be far more revealing.

How is the 'victim' responding? You may find that the victim will become increasingly reactive and nervous each time he is attacked or intimidated. This merely fuels the aggression as the bully probably feels he is approaching victory.

Specific questions regarding aggression towards humans

Is your cat displaying aggression to humans? Don't forget that your cat is armed to the teeth (literally) and the injuries that small furry creatures can inflict are worthy of a trip to Casualty. Aggression to humans, however, is a matter of degree and can range from scratches and nips to threatening behaviour and intimidation. Unless you have been there it is impossible to imagine how easy it is for a cat to stop you in your tracks just by staring at you.

If so, is it directed at one person in particular? There are often frightening parallels between cat-to-cat and cat-to-human aggression. Cats will often single out one particular individual in the household to intimidate. It is no coincidence that the person is usually female and, ironically, the one who loves the cat the most. (See Chapter 6.)

Does your cat distinguish between men and women when behaving aggressively? If your cat attacks men this is another relevant point and could indicate that he is fearful because of a previous lack of contact with men, or a frightening experience. Women usually fall victim because we are more likely to respond to aggression and back down. Most men wouldn't even notice the cat being aggressive if they were engaged in another activity at the time.

Is the aggression relevant to any particular time of day or circumstance? Your cat may only be aggressive at mealtimes or shortly afterwards. The aggression may be more apparent at dawn and dusk when the cat is naturally more active. These facts may leave you none the wiser but may be relevant to the behaviour counsellor if you go down that road.

Does the cat use teeth or claws or both? The type of 'attack' could also be relevant to distinguish between play behaviour that has gone badly wrong or genuine aggression used to get the cat away from you, or vice versa.

Does your cat vocalize/growl prior to attack, or adopt a particular posture? This will help you distinguish between genuine aggression and misdirected play or predatory behaviour. It may also enable you to see whether your cat is motivated by fear or confident anger. You will probably find it difficult to carefully assess his body language whilst screaming and diving for cover, but do your best.

Does anything specific always happen prior to or after an 'attack'?

Some nasty neurological problems or seizure disorders can result in sudden aggressive behaviour. Do not hesitate to seek professional help if you suspect this is the case.

What is the victim's reaction to the aggression? Waving your arms and screaming may be a strong reward for your cat's behaviour. Backing off may make your cat even more convinced of his omnipotence. Changing the way you react to his behaviour will often be the solution to the problem.

Is injury inflicted, and if so does it require treatment? Some acts of aggression show a degree of inhibition and are used as warnings or even boisterous play behaviour. If a bite penetrates the skin the wound is highly likely to become infected due to the host of nasty bacteria present on your cat's teeth. An infected bite requires antibiotics at the very least and should be treated extremely seriously. Some people are very vulnerable under these circumstances, particularly the elderly, heart patients or those with compromised immune systems. I strongly recommend that cases involving this degree of injury should be seen immediately by a pet behaviour counsellor recommended by your vet. Don't try DIY on this one.

Don't be downhearted if you have answered all the relevant questions and remain perplexed. This exercise is really a way of focusing the mind on what is *actually* happening in your home. Many owners describe their cat's behaviour by giving it specific interpretations such as 'He peed on my bed because he was angry with me' or 'He sprayed urine right in front of me just out of spite' or 'He hates me.' Problem behaviour doesn't arise out of anger, spite or hatred; if only life was that simple. The following chapters will enlighten you further, I hope.

CHAPTER 4

House Soiling

THROUGHOUT THE NEXT FEW CHAPTERS YOU WILL SEE THAT IT becomes impossible to view each behavioural problem in isolation. Inter-cat aggression is often associated with house soiling or urine spraying. Multi-cat households can spawn anything from urine spraying to over-grooming or even obesity. For the sake of simplicity I have addressed separate issues in separate chapters but you should really read all of them to get the bigger picture.

In the 'Feline Felons' survey (conducted into problem behaviour in 2000 with the help of a national cat magazine, *All About Cats*) 30 per cent of households experiencing problems were mopping up urine from carpets and rugs on a regular basis. There is always that overwhelming sense that your cat is being naughty and destructive out of some malevolent desire to

and next-door's cat pulls faces at me through the window.

- The litter tray is right next to the cat flap. What if a strange cat comes through when I'm using it? I can't defend myself.
- My owners use newspaper to line my tray but object when I pee on the *Sunday Telegraph* before they have read it. Isn't that unreasonable?
- Last time I used the litter tray my owner shoved a pill down my throat. I'm not going there again.
- I'm a Persian, for goodness' sake! What's a litter tray? *(Sorry, Persian owners! The good news is that I am starting to see fewer Persians for house soiling than I used to a couple of years ago so hopefully this is a good sign.)*

Cleaning soiled areas

Understanding the potential causes of house soiling is half the battle towards finding a solution. Providing the appropriate number of trays in various discreet locations indoors (irrespective of the presence of a cat flap) and using a fine grain substrate will always help these situations. Cleaning previously soiled areas is also essential but most products available are pretty ineffective if the problem has been evident for some time. Cats can potentially pass gallons of urine over a period of months and it penetrates the carpet, the underlay and the floorboards or concrete beneath. It is very invasive. Sadly there is often no alternative but to remove that section of carpet and replace it. Prior to fitting a new section it is worth treating the wood or concrete beneath to try to remove any remaining odour that may be present. Some litter additives (to reduce the smell from litter trays) contain a mineral called zeolite that is extremely absorbent. This can often collect any remaining gas that may be lurking in the floor if it is sprinkled liberally over the area and vacuumed away forty-eight hours later. Whilst it's working this product looks and smells like an open-plan litter tray so it's important to keep your cats away. Other new cleaning products are emerging that are proving extremely effective; the staff at your veterinary surgery may be able to recommend a product that they have tested themselves with good results.

Feline Idiopathic Cystitis

Over the years I have seen hundreds of cases and a worrying pattern has emerged. More and more cats are suffering from urinary tract disease that is directly influencing their toilet habits and behaviour. Veterinary research has shown that a large percentage of all cases of urinary tract disease are *not* directly related to crystal or bladder stone formation or bacterial infection as was previously believed. A newly recognized condition is referred to as Feline Idiopathic Cystitis (FIC) but the actual cause of the problem remains unknown. It tends to be more common in middle-aged, sedentary and overweight cats that live in multi-cat households. Symptoms include

- straining
- difficulty in passing urine
- blood in the urine
- inability to pass urine
- urination in inappropriate places

Occasionally complete urethral obstruction can occur in male cats and this requires emergency treatment to prevent kidney failure. Other reported behavioural changes include over-grooming around the lower abdomen (possibly in response to the pain), reluctance to be handled and aggression towards people or other cats in the household. This is an extremely distressing problem for both the cat and the owner and a great deal of research is being conducted to try to discover more about the condition and the underlying cause. One theory shows how a complicated relationship between the nerve supply to the bladder, the urine and the lining of the bladder may be at the root of the problem. It is, however, well established that stress is a major trigger.

A number of treatments are currently being prescribed by veterinary surgeons including dietary supplements to promote repair of the lining to the bladder and even, in severe cases, potent anti-depressant drugs. However, the latest research indicates that the most effective treatment for this condition involves altering the cat's lifestyle by changing the diet to increase water intake and reducing stress.

It is often hard to imagine that our beloved felines are stressed but it's important to understand that their concerns in life are very different from our own. Stress can be triggered by changes in weather, diet or routine, or by the introduction of new people to the household. However, by far the most potentially stressful element in any cat's life is . . . other cats! (See Chapter 7.)

It is extremely important that owners seek veterinary advice when their cat shows any signs of urinary tract disease. However, this is one of the many conditions that require both medical and behavioural treatment. Owners of cats with Feline Idiopathic Cystitis can make a real difference to their pets' well-being if they observe the following guidelines.

Reduce stress

- Try to limit the number of cats within the household to a

reasonable level given the size of the house. There is no rigid formula for this but the number of cats within the territory should also be taken into consideration to prevent further overcrowding.

- Ensure that a sufficient abundance of resources is available within the home to prevent competition. These resources include beds, litter trays, high resting places, private areas, water containers, food bowls, toys and attention from the owner.

Encourage more frequent urination

- Provide the appropriate number of litter trays within the home, irrespective of whether the cats have access to outdoors. The number of trays provided should be 'one per cat plus one' in different locations. The places chosen should be in quiet corners away from thoroughfares. Trays should also be positioned away from the cats' food.
- The litter trays should contain a fine grain substrate and any soiling should be removed daily.
- Many cats with Feline Idiopathic Cystitis are overweight so increased activity and play, together with a reduction in the food provided, will help to reduce weight. More active cats tend to urinate more frequently.
- Previously soiled areas (e.g. carpets, sofa cushions) should be cleaned thoroughly to prevent any further association with an appropriate toilet location. Heavily soiled areas of carpet may ultimately need replacing.

Increase fluid intake

- Water intake should be increased by feeding a wet diet. Small amounts of water can also be added to the food.
- Water should always be provided in alternative locations to

the food and in a number of places within the home.

- Install pet drinking fountains or water features to provide a source of moving water.
- Provide plastic tumblers on bedside tables or window sills.
- Replace the water frequently and use filtered or pre-boiled water.
- Flavour the water using a pinch of gravy granules, cat milk, fish or chicken stock.
- Ensure bowls are available outside to collect rain water.

We all live such busy lives that often the art of cat behaviour counselling (or even the art of playing cat detective) is much easier when someone from outside comes in with a totally objective view of things. Often clients will remark that everything I have told them to do and the reasoning behind it is pure common sense. I would agree to a point, but if it is just common sense why do cat owners need me in the first place? The answer is that sometimes owners are just too close to the problem to see the wood for the trees.

The next case illustrates this point. I hope it will also strike a chord with many owners experiencing a soiling problem indoors and help them see that sometimes the easiest solutions can result in a miraculous cure.

Bonnie – the case of the flatulent cat

Bonnie and Clyde were a couple of one-year-old black and white moggies living with Amanda, her husband and two young children in a happy family home in Surrey. I remember this case vividly because of Bonnie! The two cats were brother and sister and very similar apart from certain markings around their faces. It's extremely important for me to recognize individuals once I've been introduced; anything less would look as if I didn't care. I usually find one trait that I can identify at a glance. Bonnie had a very distinctive black splodge on her face around her nose and mouth that, from a particular angle, looked like a little protective mask. I remember thinking how fitting this was as she had the worst flatulence of any cat I have ever met. She had a slightly distended abdomen and as she ran around the living room chasing a toy she left behind her the most malodorous vapours. It's a shame poor Clyde didn't have the mask! Amanda apologized and said that she had suffered from a rather grumbling bowel and intermittent diarrhoea since she was a kitten. A full veterinary investigation had taken place apparently and nothing specific had been diagnosed.

Both cats had access to outdoors via a cat flap in the door of the conservatory but Amanda kept them in at night with a litter tray because her previous cat had been killed on the road during the night and she didn't want to tempt fate with her two new ones. She had rather hoped that they would toilet outside during the day and hold everything overnight so that she could get rid of the tray. Crawling toddlers and litter trays don't always go too well together and they require a huge degree of vigilance from already busy mothers. Unfortunately both Bonnie and Clyde had been urinating and defecating in various places throughout the house ever since they had arrived. It was dismissed initially as varying degrees of 'accident' but the problem was definitely getting worse. Amanda had positioned their litter tray in the middle of the conservatory floor, added another one next to it, tried various litters and even mixed litter with soil, but nothing seemed to resolve the problem. It was frustrating for her to see both cats choose areas round the edges of the conservatory or in the pot plants but avoid the

trays like the plague. Amanda's husband had put his foot down and given her an ultimatum. The cats clean up their act or go!

The facts of the case

Here are the important points that I gleaned during my visit.

- Bonnie had a long-term bowel problem causing a degree of 'intestinal hurry' when she needed to pass faeces.
- The litter trays were positioned in the middle of the floor, side by side.
- The trays were the same size as the original small one used when they were tiny kittens.
- Bonnie and Clyde were both bullied outside by more assertive cats.
- Amanda was a very busy young mum and she was occasionally a bit lax in her tray cleaning.

Any problem associated with the bladder or bowel can

potentially disrupt acceptable toilet habits. Bonnie had always had this rumbling belly and when she needed to go it was an urgent requirement. She might have had unpleasant or painful associations with the existing litter facilities, or just got into the habit of soiling in other places in response to the urgency to go. Either way it was imperative that we looked at her diet and checked with the veterinary surgeon that everything possible was being done to settle her stomach and bowel.

Amanda had done her best by providing the litter trays but her choice of location was not very appealing. Placing a tiny tray in the centre of the conservatory meant it could be viewed from every angle by all the nasty cats outside the semicircular room, and offered no protection whatsoever. Placing an equally small tray alongside didn't make any difference at all to their appeal. Owners can get very frustrated and switch litter materials with incredible speed. This doesn't help either and just makes the whole area one to avoid at all costs, particularly if the few pees or poos that actually end up there seem to remain for an eternity!

The other very significant element was the bullying going on outside. Neither Bonnie nor Clyde was particularly keen on putting their lives on the line by eliminating outdoors. The flower beds were at the bottom of the garden and to get there and back without being beaten up was hard enough without trying to poo in between. The indoor facilities were therefore of paramount importance and they just weren't up to scratch. Bonnie and Clyde therefore had to make it up as they went along.

The behaviour programme

I devised a plan that was specifically tailored for Amanda. It is pointless giving a busy mum a complicated programme to resolve a problem. With the best will in the world it will not be followed and the problem will not go away, and that will be my fault. So I went for the minimalistic approach with the following:

- Amanda was to take a faecal sample (beautifully provided by Bonnie and collected by me during my visit!) to her vet for analysis to make absolutely sure there was nothing sinister going on. I also spoke to the referring vet and she suggested worming and the use of probiotics to rebalance Bonnie's natural gut bacteria after such a long period of disruption.

- Amanda changed Bonnie and Clyde's food gradually over a period of seven days to a complete dry diet. They used to receive rather rich sachets of wet food which appeared to pass through Bonnie and leave her in exactly the same form.

- I asked Amanda to buy two large open litter trays and place one in a corner of the conservatory, protected from sight from the house by a palm tree, and the other in the opposite corner protected on two sides by a pair of large umbrella plants.

- Amanda was given a strict cleaning regime for these trays and asked to use a fine grain sand-like substrate.

- There was a small potential toilet area outside in a raised bed near the cat flap that I felt could be a safe outdoor facility so Amanda's husband did his bit and weeded the area, mixing the soil with peat and sand to make it light and potentially attractive to Bonnie and Clyde.

- I also gave Amanda a cleaning regime for the soiled areas that would remove any residual odour.

The outcome

That was all I asked her to do and the results were dramatic. Bonnie responded well to her new diet and her stools became firm within a matter of days. The consistency of Bonnie's bowel motion became quite a source of delight for both Amanda and me (I know, I really should get out more). Both Bonnie and Clyde took to the new litter trays immediately and Clyde was even observed playing around in the newly prepared soil outside. After Amanda changed everything there was only one accident and she admitted that she had been less than careful over the cleaning of the trays that particular weekend because she had been entertaining guests.

Bonnie's tummy was no longer distended and her 'silent but violent' emissions were a thing of the past. I must admit I was happier for the family about this particular achievement than I was about their pets' new-found toilet habits!

Smartie – the case of the frightened pensioner

Treacle and Smartie (brother and sister) were two eleven-year-old black and white short-haired cats. I met them and their owner Richard three months after they moved to their new home. Life had been rather difficult for them recently. Two years previously Richard's home had flooded, requiring extensive repairs spanning many months. He had managed to remain in the property by utilizing the first floor and the cats had adapted well to the changed nature of their environment. They continued to come and go through the cat flap downstairs but, once in, remained permanently upstairs. The repairs were completed and life returned to normal but in the meantime Smartie (the female) had started to occasionally urinate and defecate in Richard's bedroom. He had never provided indoor litter trays and he couldn't understand why Smartie was suddenly deciding to do this. She was still using the cat flap to go outside and it was only an intermittent problem so he didn't really make an issue of it. But he remembered there had been a problem at the same time with a stray cat which came through the cat flap and ate his cats' food.

Time passed and Richard sold his house and moved to his new home. Both cats took a while to settle in but they soon used the cat flap and explored their garden. Richard had temporarily supplied a litter tray in the utility room for their period of confinement but he removed it as soon as they discovered the great outdoors. Shortly afterwards the soiling problem returned and Smartie urinated and defecated regularly in various corners of the house, both upstairs and down. There were probably a total of five locations that Smartie would return to and use as lavatories. Richard was keen to find a solution and his vet recommended that he place a litter tray in the utility room again. This did coincide with an improvement, but the problem still continued at a level that required more serious assistance.

The facts of the case

Treacle was out when I arrived but Smartie was asleep on the sofa in the living room. Richard discussed the problem with me and the following points were particularly significant.

- Smartie had become more dependent and affectionate recently.
- Smartie was definitely going outdoors less than she used to before the flood in Richard's previous home.

- She chose a number of places to return to repeatedly when she soiled.
- She often cried and appeared agitated before she soiled.
- The provision of a litter tray reduced the incidence slightly.
- There had been an invading cat in their previous home.
- There were a number of cats in the neighbourhood in their new home and Richard had fitted an exclusive entry electronic cat flap to keep them out.

Cats can be extremely threatened by other cats in the locality, particularly if they have a history of invading their homes. Smartie and her brother had always urinated and defecated outdoors but, in the presence of threatening cats, this activity is often fraught with danger. Some cats just find it incredibly stressful and look to the comparative safety of their homes for a suitable loo. If there are no litter facilities they will often find a private corner in a room where they feel secure. As cats get older they become less confident in their ability to defend themselves. They often go out less and stay nearer to their home when they do venture outdoors. They become more dependent on their owners to provide comfort and security. Older cats thrive on routine so a house move would have been potentially quite distressing for both Treacle and Smartie. Their individual personalities dictated how they responded to this challenge and it was apparent that Treacle was rather more robust emotionally. Add all these issues together and it becomes clear that Smartie was becoming more and more concerned about eliminating outside. The vocalization before she urinated or defecated indoors would indicate there was some stress involved in the decision-making.

Richard had the right idea when he provided an indoor tray. Unfortunately, in Smartie's mind, there were still five other alternative sites to choose from so her use of this litter box was only intermittent. Once a cat has urinated or defecated in a particular area it will continue to return as a response to location, residual smell or texture underfoot. A habit soon develops and the place becomes a lavatory as far as the cat is concerned.

The behaviour programme

It is very difficult to encourage cats to eliminate outdoors if they perceive there is a threat from other cats, particularly as they get older. Richard would have to accept that litter trays were going to be a permanent addition to his home if he wanted to stop Smartie from using his carpets. I suggested an easy solution.

- All five areas of soiled carpet were cleaned thoroughly to remove any residual odour.
- Richard continued to provide the tray in the utility room but he would also position two others to give Smartie a choice.
- A covered tray, with a wide opening at the front, was located in a quiet area of the hallway in a recess under the staircase. A further open tray was located upstairs in the corner of a spare bedroom where Smartie liked to rest occasionally.
- All three trays were filled with a fine sand-like substrate and no polythene or newspaper was used to line them.
- All trays were checked twice a day and any urine or faeces removed. Twice a week the trays were emptied completely, washed and replenished with fresh litter.
- Richard was asked to play gentle games with Smartie. I almost always use play therapy because I have found it to

be very helpful. Many owners believe that it isn't necessary or even pleasurable to play with older cats but most love it if encouraged in the right way. Smartie and Treacle took to the games with great enthusiasm and they would play every evening. Richard also found this surprisingly relaxing!

- I asked Richard to provide a number of warm soft beds in various locations together with a tall modular scratching post. Smartie liked to rest in high places and there was a perfect area for a 'cat gym' near a radiator in the large kitchen/diner overlooking the garden.
- Doors to the two bedrooms where soiling had occurred were kept shut to prevent Smartie's being tempted to use the familiar locations. The other three areas were in the hallway and living room so these were disguised by moving furniture over the spots so that Smartie couldn't physically reach them.

The outcome

Almost immediately the soiling stopped and Smartie used all three trays regularly. Richard was also delighted to report that she appeared livelier and more interested in life, and he even found her playing with a rolled-up piece of paper in the hallway all by herself! I recommended that the three trays remained indefinitely.

Suki – the case of the old cat that learned a new trick

Who says you can't teach old cats new tricks? Suki and Holly were thirteen-year-old sisters, both beautiful long-haired tortoiseshells. One was dark brindle and the other had huge splodges of ginger, black and white. They lived with Edward, a very distinguished and rather lovely middle-aged gentleman, in a third-storey converted flat in West London. Sadly his marriage had ended some five years previously and he had moved out of the family home, taking his two cats with him.

Suki and Holly had always enjoyed an active outdoor life in their previous home but they were confined in their new one apart from access to an attractive small roof terrace. Edward felt sure that they would adjust to their new surroundings. He did all the right things and took his responsibility very seriously. He positioned two deep litter trays in an eaves cupboard in a small room that he converted to an office. A cat flap was installed in the cupboard door, providing a private and discreet bathroom for the two ladies in his life. He was a busy man so he felt the best option was to pour an entire bag of litter in each tray and then replace the whole thing on Sundays. Perfect low-maintenance solution!

Suki seemed to think otherwise because immediately she arrived in her new home she started to urinate and defecate in

just about every corner she could find in the flat. This exasperated Edward but he cleaned up after her until she appeared to settle on one area behind the sofa in the living room as the most suitable spot. He continued to clean as part of his busy daily routine but that carpet soon had to go and the eventual compromise was expensive laminate and a huge sheet of thick polythene upon which Suki toileted relatively consistently.

I'm not entirely sure why Edward waited five years to call me. I think we all know what it's like when time plays tricks with you and weeks, months and years pass by in a complete blur. Edward's life had changed dramatically and he was adjusting himself to his new solitary lifestyle and increased pressure at work. Five years can go very quickly.

The facts of the case

I liked Edward and I could see that he didn't really want the whys and hows. We were after solutions so I launched straight into a plan to put things right. I may not have gone into great detail with Edward but here are the important points.

- Suki and Holly had previously eliminated outdoors.
- The litter trays in their new home were deep-sided, polythene-lined and filled with a non-clumping litter.
- The litter trays were located within a small cupboard with a cat flap entrance.
- The cats had to face a further assault course to weave in and out of all the papers and files on the floor in Edward's office to get to the cupboard in the first place.
- Suki and Holly were not the best of friends, preferring to keep a reasonable distance apart at most times.
- The litter was changed once a week.

- Suki was now only eliminating in one location in the living room apart from the occasional poo in the litter tray, usually at the start of the week.

Life had changed a great deal when Suki and Holly moved to their new home. Suddenly access outdoors wasn't an option (they did indulge in the occasional precarious balancing act over the roofs of neighbouring properties) and every need had to be fulfilled indoors. Suki and Holly had always tolerated each other because it was easy to put a great distance between them. That wasn't quite so simple in their new home and tensions appeared between the sisters. They had to eat together and, worst of all, toilet together with no possibility of choosing alternative sites (in theory). To make matters even worse the provided facilities were difficult to get to and there were plenty of opportunities for a mischievous Holly to block Suki's access just for the hell of it. Edward really wasn't aware quite how many objections Suki potentially had to his convenient toilet: inadequate cleaning, polythene liners that got caught in her claws and threw litter in her face, sharing with another cat and no ventilation in the cupboard, to name but a few.

Suki took the only action available to her: she started to look for her own convenient lavatory. Eventually settling on a discreet area behind the sofa, she then had a place she could call her own. The polythene sheet that Edward put over the area merely acted as a trigger and reassurance that she was in the right place.

The behaviour programme

I knew I had to devise a plan that was easily achieved by a very busy man with loads of other things on his mind. He really was looking for a quick fix but he was smart enough to know that

that was a little unrealistic. I felt the following would do the trick.

- Edward had found the stench coming from the litter cupboard rather overwhelming when he worked in his office so he installed a number of plug-in air fresheners. In this job there is only one smell worse than stale urine and that is plug-in air fresheners trying extremely hard to compete. They had to go, for the sake of everyone; these are strong smells for cats that are pumped out right in their faces.
- Edward had to go shopping for two additional litter trays. I suggested the type that maybe wasn't so high-sided (I think he was actually using large washing-up bowls!) but was large enough to allow a cat to turn round and manoeuvre with ease.
- One of the new trays was positioned in a spare bedroom and the second placed behind the sofa in the living room on top of the polythene sheet.
- One thing I would not compromise on was litter hygiene. This was obviously an issue and I reminded Edward that he was performing an even more laborious chore when he cleaned up after Suki every day. We replaced the litter with a fine-grain clumping type and Edward used just enough litter to provide a depth of one inch in each tray.
- Every day he removed clumps and solids and topped up with fresh whilst keeping a daily diary of what he found in each location.
- I recommended a zeolite powder (available in good pet shops as a litter deodorant) to line the trays to reduce odour, particularly as one was located in his living room.
- I also felt that both cats needed their separate private areas

and I asked Edward to make sure there were plenty of places for them to take 'time out' from each other.

The outcome

Our long-term plan was to dispense with the polythene but, as using it was such an entrenched behaviour in Suki, it had to be done gradually.

Edward worked hard and sent me the most comprehensive diary of his cats' bowel and bladder habits. The most exciting thing was that they were both using all the trays and Suki only occasionally used the polythene. After four weeks Edward started cutting strips off alternate edges of the polythene one inch at a time. This was a big sheet of polythene but I'm glad I advised him to do it this way. After about three months the polythene was reduced to the size of the tray and Suki was directing everything into the tray.

Edward and I did disagree about something, though. He desperately wanted to remove some of the trays but I knew this would be a slippery slope to problems again in the future. The cats were much more relaxed and the new toilet arrangements were largely responsible. My only possible suggestion was to move the tray behind the sofa one inch at a time towards the hallway and relocate it over a period of several months. We eventually agreed on landscaping the area behind the sofa using pot plants to prevent any visitors from seeing the tray. This seemed like a reasonable compromise and I hope that Edward, Suki and Holly are still happily enjoying their life together.

CHAPTER 5

Urine Spraying

URINE SPRAYING IS A PERFECTLY NORMAL FELINE BEHAVIOUR
and it represents an important means of communication within
a cat's territory. The behaviour, when observed, is quite un-
mistakable. The cat will approach an object or vertical surface
and often sniff intently. He may even turn away slightly from the
point of interest with an open-mouthed grimace and a faraway
look. This shows us how interesting the smell must be as the cat
draws the exciting aroma into the second organ of scent that a
cat possesses in the roof of its mouth (called the Jacobsen's
organ). Once all the necessary information has been gleaned
from the 'tasted' smell the cat will turn round, raise its tail
vertically and tread with its back paws. As the tail quivers a
small jet of pungent urine hits the surface. Hey presto! A unique
scent and visual mark is deposited for all to see.

As cats appear to be able to differentiate between sprayed and squatted urine it is believed that the anal glands may secrete when urine is sprayed to produce the oily, viscous liquid found on the skirting boards of less fortunate cat owners. There should be, in theory, no need to spray urine indoors if it is perceived as the cat's core area or den. Safety and security in this context should be paramount. If the individual develops a sense of insecurity and becomes stressed then he has limited ways of expressing this vulnerability. So he uses a natural behaviour (urine spraying) usually employed in situations of conflict.

The jury is still out regarding the definitive reasons why the behaviour is so necessary but it appears to be most relevant to the sprayer itself. Any cat is capable of spraying urine, male, female, neutered or entire, although it is most common in the intact male. Sexually active cats will spray urine that is laden with pheromones to indicate their readiness for mating. Neutered cats will spray on fences and bushes, for example, in areas of high cat density as part of their daily routine. Despite the fact that urine spraying can be utilized to relieve all sorts of weird and wonderful emotions in certain rather complex individuals it can safely be said that it is usually 'a cat thing' and another feline is at the root of the problem.

One interesting statistic gleaned from the 'Feline Felons' survey (see page 51) reinforced my feelings about cats' being acutely aware of social overcrowding. The results showed that the incidence of urine spraying indoors increased in proportion to the number of cats in the home, from 17 per cent of single cat households to 86 per cent of those with seven or more cats. Every house has that 'one too many' cats threshold. Two may be a crowd in homes containing particularly intolerant individuals whereas six or seven may be the magic number in others. A great deal of luck is required, together with the right environment, to prevent all hell's breaking loose (see Chapter 7 for more information about the multi-cat household). Indoor urine spraying is often just one symptom of a generally turbulent and disrupted existence. Most cases are combined with other complications including excessive scratching within the house, anxious individuals in the group and, worse still, inappropriate urination or defecation indoors.

Urine spraying is not the only marking behaviour that cats indulge in; their sense of smell is the most powerful of all their senses and scent is deposited daily by various means.

Rubbing/chinning

Cats have sebaceous glands around lips, chin, head and base of the tail. These secrete scent that can be used either to rub against other cats within a social group to identify members of the same gang or to deposit on objects around the home and outer territory to ensure that things smell familiar. You will probably notice that your cat will raise his bottom if you tickle him on top of the base of his tail or push against your hand if you stroke him on his cheek. He's probably just grateful that you're going to help him spread his scent around.

Middening

When cats feel threatened within their home territory they may occasionally deposit faeces in prominent locations and on strategic pathways as a very strong signal to all and sundry. It is sometimes easy to confuse this marking gesture with a case of inappropriate defecation when the problem is not invading cats but merely a dirty litter tray.

Scratching

Scratching performs two important functions. It is a common misconception that cats sharpen their claws by scratching. What actually happens is that the worn outer husk of the claw is detached, against a resistant material, revealing a sharp new surface underneath. Scratching also exercises the muscles of the forelimbs. In addition, it is used as a form of marking. Scent and sweat glands in between the pads of the paws mix their secretions to produce a unique smell. When claws are scraped down a surface the scent is deposited and the combination of the scratch mark, the discarded claw husks and the smell provides a strong visual and scent message to other cats.

Scratching is the marking behaviour that causes the most damage in the average home. It's probably worth focusing on this for a moment because, if you are experiencing problems, there are ways of keeping your furniture, carpets and wallpaper intact.

Where do cats scratch?

Cats will choose a variety of surfaces both vertical and horizontal depending on their individual preferences. Cats often like to stretch and scratch on a horizontal surface when they first wake up. If the surface is vertical the cat will usually extend itself to full stretch and then rhythmically scratch, alternating between the fore paws. Others will scratch by lying down and pulling their body along the floor. The surfaces chosen are usually fixed and non-yielding to resist the force exerted by the cat as it scratches.

Scratching outdoors

Trees, fence posts, sheds and wooden gates in strategically important locations will all show signs of marking behaviour in a cat-populated area. Similar surfaces will also be utilized for claw maintenance. Unvarnished woods and tree bark are the most natural surfaces to scratch as they provide a perfect level of resistance to the action and show a strong visual sign when used regularly.

Scratching indoors

Your cat may have limited or no access to outdoors. Alternatively he may choose to spend more time in the comfort and safety of the home and just feel more relaxed about maintaining his claws in a secure environment! Scratching can also be used as a precursor for play or even as an attention-seeking tool by the more manipulative and social individuals.

Popular substrates indoors include soft woods (e.g. pine), fabrics, textured wallpaper and carpet. Popular locations include door frames, furniture and stairs. Cats will often scratch vigorously in the presence of their owners or other cats as a sign of territorial confidence. However, if the scratched locations are widespread throughout the home, particularly around doorways and windows, then it is likely that the cat is signalling a general sense of insecurity. Whether the scratching represents claw maintenance, marking or both depends on the dynamics of the feline household, the pattern of locations and various other factors. Your challenge is to work out what's going on in your home.

There is one important thing to remember. If attractive

scratching posts or areas are not provided indoors it is likely that damage will occur to furniture, wallpaper or carpet. You have been warned.

Prevention is better than cure

If your cat is owned from a kitten it is important that he becomes accustomed to handling and restraint at an early age. If a cat becomes used to claw trimming as a kitten then he will tolerate it well as an adult and damage to furniture will be prevented. This may be a beneficial strategy for cats kept exclusively indoors.

Commercially available scratching posts range from a basic single upright structure with a heavy base to an elaborate floor-to-ceiling modular unit that provides many opportunities for play, exercise and resting as well as a variety of surfaces to scratch. In multi-cat households it is probably sensible to provide one scratching post per cat (plus an additional one for choice) positioned in different locations. The choice of design depends then on budget and space available.

If space is an issue then scratching panels can be fixed to walls, either using home-made or commercially available products.

- Sections of carpet can be attached to walls using double-sided carpet tape and wooden batons at the top and bottom (fixed with rawlplugs and screws) for added security. The carpet chosen to provide a suitable surface for scratching should be a loop-weave to offer the appropriate degree of resistance. It is also essential that it is positioned to allow the cat to scratch at full stretch (remember that kittens grow very quickly so full stretch for them will not be high enough).
- Commercially available panels of sisal twine can also be attached to walls to create a similar area.

Avoid those scratching products that are too lightweight to resist scratching or cannot be fixed to rigid surfaces. These will not be favoured by cats due to the lack of resistance when used. And don't worry – although many commercially available scratching posts are covered with carpet there is no evidence that the cat's scratching habits will generalize to other areas of carpet within the home once the post is used regularly.

Introducing your cat to a scratching post

It is important initially that the post, scratching panel or modular 'cataerobic' centre is located in an area your cat frequents on a regular basis. Placing it somewhere convenient for you but not visited by your cat will guarantee that it is ignored! As cats often scratch after a period of sleep it may be useful to place the post near a favourite bed.

If your cat is used to handling and manipulation of the legs and paws then it may be sufficient to gently hold him against the new post and gently scrape his paws down the surface whilst offering praise. However, many cats will be suspicious of an overzealous owner if they are actively encouraged to explore the new item. There are better ways to make the scratching post irresistible.

Some commercially available posts are impregnated with catnip. This herb is extremely attractive to many adult cats and its presence will often draw attention without much effort. Once the cat has approached the scratching post a simple predatory-type game (involving a feather attached to a piece of string, for example) around the base will encourage him to give chase and his claws will then make contact with the surfaces of the post. Often this will be sufficient to encourage further visits. If the scratching post has several levels then placing tasty dry food on the modular surfaces may encourage the less playful cat to investigate.

Damage to furniture from scratching

Prevention certainly is better than cure, but if you are in need of a cure here are some suggestions.

- If a particular surface or object is being damaged it is important to provide an acceptable alternative that offers a similar experience when used. For example, if your cat is scratching textured wallpaper at a certain height it is essential that the alternative scratching area is vertical with similar texture and striations that allows him to stretch to the same level.

- It's important to remember that your cat is not doing this just to be naughty. If the motivation is claw maintenance then you are merely punishing a natural behaviour (very confusing for the cat). If your cat is scratching excessively due to anxiety and insecurity then punishment will add to his distress and probably make the situation worse.

- If scratching has damaged furniture, it is possible to deter your cat from future visits to the same location. Low-tack double-sided adhesive tape* can be stuck over the area and this will provide an unpleasant (but not dangerous) experience when your cat next scratches there. It is essential to ensure that the tape is not too sticky since it could damage paws and fabric. This method can be employed once there are acceptable scratching posts nearby to use as an alternative. Commercially available double-sided adhesive

---

* The adhesive on the tape will attract atmospheric dust and fibres so it may be necessary to place a fresh strip over the original on a daily basis if the cat is persistent with his attempts to scratch.

sheets can be purchased from some household cleaning suppliers specifically for this purpose.

- If wooden furniture, door frames or banisters have been damaged by scratching it is important to remove all traces of the scratch marks by rubbing down with a fine sand-paper and treating the area with a thick layer of furniture polish once the surface is smooth again. Suitable posts or scratching panels should be located nearby. If the area is not ideal for a free-standing scratching post on a permanent basis then it can be relocated slowly (an inch at a time!) to a more convenient position once it is being used regularly.

- Many cats target carpet on the lower step of staircases and scratch horizontally whilst lying down. If this is occurring in your house then place low-tack double-sided adhesive tape over the damaged areas (warn the family not to tread on it) and provide a scratching area nearby. If your cat grips the stair on each side of the right angle, providing both vertical and horizontal scratching surfaces, it is im-portant that the alternative offers the same opportunity. For example a breeze block covered in carpet will be heavy enough to resist the pull of the scratching action, can be used for both vertical and horizontal scratching and is easily located nearby.

- If your wallpaper is being damaged then thin sheets of Perspex can be cut to size and fitted over the damaged area, using screws and rawlplugs if appropriate. This surface will be unattractive to scratch since it is smooth and it is also easily cleaned to remove any scent deposits. Double-sided adhesive tape can also be used over the affected area if the wallpaper is sufficiently damaged to require replacing. Whichever deterrent is used it is essential to provide a vertical scratching panel of a similar height nearby.

Scratching as a marking behaviour

If your cat has chosen a variety of locations to scratch and these are regularly targeted it's important to rule out territorial marking as a motivation. Scratching may be anxiety-related if the following is the case.

- It is widespread.
- It is present in a multi-cat household.
- It is present within a home in a densely cat-populated territory.
- There have been major changes within the home.

There are often tensions within multi-cat households or territories that are not easily identified by the owners. The solution to territorial marking lies in identifying the problem and the cause of the individual cat's stress (see Chapter 7). Once a potential cause has been established it may be possible to decrease the cat's anxiety by providing additional resources to prevent competition between members of the group. Making environmental changes within the home will also increase the cat's feelings of security and safety; the following suggestions may be useful.

- Provide a number of high resting places and secure hiding places in different rooms (one per cat plus one is always a good formula to follow).
- Provide sufficient indoor litter facilities in different locations if the cats have limited access to outdoors (one per cat plus one).
- Increase interactive play sessions.
- Provide additional food bowls elsewhere in the house.
- Provide additional water bowls elsewhere in the house.

- Ensure there are plentiful sleeping areas and beds, even if they are cardboard boxes containing old jumpers.
- Cover the cat flap with a solid panel, on both sides of the door, and give the cat access outdoors on demand.
- Provide an electronically controlled exclusive-access cat flap.

Deterrents to further scratching

There are various commercial deterrents on the market that can be sprayed on the damaged area to prevent further approaches from your cat and his claws. In my experience these products emit an odour that is also highly offensive to humans and the spray needs to be regularly reapplied to be effective. You may end up with a pristine sofa that you are unable to enjoy because it smells so bad. Other deterrents that may prove more useful include

- tin foil that can be used as an alternative to double-sided tape
- small vinyl pads called Soft Paws that can be glued over the cat's claws by a veterinary surgeon and will remain in place for 6–10 weeks; scratching will still take place but damage will no longer occur and the cat can be retrained to more acceptable areas
- synthetic or natural feline facial pheromones that can be sprayed over the area that is being damaged by scratching (see Chapter 2)

❖　❖　❖

Let's focus once more on the most distressing of all the marking behaviour, urine spraying. If you haven't lived with this problem you cannot imagine how disruptive it can be. Urine spraying can often be an intermittent problem and many households tolerate it for years because, in between episodes, they keep hoping that it has gone away. Sadly it comes back in most cases so don't leave it any longer. Here are a couple of case histories that you may be able to relate to if you are very unlucky.

Jake – the case of the battle-fatigued cat

Jake was a slight (and somewhat feminine) ginger moggy with four symmetrical white socks. He was about five years old when I met him, and his owners, Laura and Chris, had been having a few nightmares with him since moving to their new home a year before. In their previous house Jake had been a gentle and placid character; he had spent a lot of time outdoors and returned home for a fuss and a cuddle when it suited him. He seemed to be the perfect cat for a working couple – independent yet loving when his owners returned after a hard day. When Laura and Chris moved they were keen to do the right thing with Jake and they tried to keep him indoors for three weeks to acclimatize him to his new surroundings. This proved extremely difficult as he was desperate to explore and they soon relented and allowed him to investigate his territory.

Little did Jake know that he was entering a battle zone teeming with warring factions drawn from the meanest feline feral fraternities. Huge beasts with broad chests and squinty eyes patrolled the territory and the arrival of Jake represented nothing more than a minor inconvenience. He was severely beaten up on his first excursion and continued to be bitten, scratched and squashed on a regular basis. Laura would often lie awake at night and hear the screams of vicious cat fights and wonder what was being inflicted on poor Jake. Chris was far more confident about Jake's ability to defend himself and felt this was merely a period of adjustment as Jake found his paws in his new home ground. They had been allowing Jake access to outdoors via a bathroom window but Chris thought the time was right to fit a cat flap to accommodate unlimited comings and goings.

Time passed and Laura and Chris noticed that Jake was finding a thousand excuses not to go outside. He was sitting by the window in the living room and 'asking' to go out from the front of the house. Since the property was one of a long terrace it was clear to both Laura and Chris that a different group of cats populated this area. Maybe Jake just didn't like the rough lot out the back? So they allowed him to go out through the living-room window, but even this was a short-lived activity. Jake once again retreated indoors or sat disconsolately on the front doorstep.

After a couple of months Laura returned home to the sinister aroma of cat pee in the living room. This confused both Laura and Chris for a number of days until Jake sauntered into the room, backed up to a chest of drawers and sprayed a fine jet of urine all over the front. After the initial shock they shouted at Jake for this appalling act of vandalism and chased him out of the room. He was subsequently barred from the living room but he continued his 'dirty protest' against the walls in the hallway and the front door. Laura and Chris would always shout if they saw him do it but this didn't seem to be a sufficient deterrent. Poor Jake became agitated and restless and he would often be found pacing around the ground floor and crying constantly.

During the course of the consultation Jake didn't settle. He took little interest in the toys in my magic bag* and spent long periods looking out of the windows. We were in the living room so both Laura and Chris were tightly coiled springs waiting to leap up if Jake showed even the remotest interest in a wall or a piece of furniture. I had a deep sense that Jake was guarding his property against an enemy far greater than himself. That must be pretty scary and comparable with a child's being relentlessly bullied at school or an elderly lady's being terrorized by vandals. Hideous.

The facts of the case

My job relies on information gathering and history taking and this, together with an almost intuitive understanding of the patient's emotional state, helps me to get to the bottom of the mystery in each case. I had been told a number of significant facts during my visit:

- Jake had always been provided with an indoor litter tray but he never used it in their old house. Now he was using the tray often.
- Jake was spraying urine in relatively large quantities.
- Jake started spraying when the cat flap was fitted.
- Jake paced and vocalized prior to spraying.
- He spent a lot of time upstairs and never sprayed urine there.

* My magic bag is nothing of the sort really. It is merely a rather battered briefcase that contains the tools of my trade. My patients, just because they are my patients, tend to be rather suspicious of strangers or reluctant to interact because they are mentally preoccupied with their worries. I need to see the cat 'underneath' the problem behaviour and play is a good way to remove inhibitions. Inside my bag I have accumulated the most popular toys that I have tested over the years and found to be irresistible to most, if not all, cats. There are tiny mice made of real fur, long bootlaces, a wonderful fishing rod toy called a Cat Charmer made of a soft multi-coloured fabric, scraps of rabbit fur, three home-made 'Octopus' toys, a couple of valerian tea bags and a packet of catnip. See Chapter 11 for details of how you too can have the ultimate collection of exciting toys without spending a fortune!

• He started to demand entry and exit through the front door.

This was my conclusion. Laura and Chris had moved into an area with a high population of feral cats. The physical appearance of the males (broad chests and thick jowls) indicated that they were entire and this would mean that the colony was probably actively breeding and expanding. Whenever a new cat moves into an established territory there is a need to fight (or at the very least agree by mutual consent) for rights of passage. Poor Jake with his rather camp demeanour just didn't stand a chance. Suddenly he was a prisoner in his own home. Well, that is until Chris fitted the cat flap! Everyone who consults me soon learns that I am not a great fan of the cat flap (see the questionnaire in Chapter 3) and, in this case, it merely represented a breach in the defences and an opportunity for the enemy to invade Jake's home. Jake had done the sensible thing and tried to establish territory elsewhere. However, he had obviously met resistance there also and his confidence would not have been at an all time high by then. He was utterly defeated so he withdrew to the comparative safety of the upstairs (cats often go up when in danger).

The use of the indoor litter tray was also significant. Here was a cat who preferred to eliminate outdoors. In his new environment this would have been a dangerous habit so he turned to the relative safety of the indoor tray. Two problems here: the tray was located in the downstairs bathroom (which was perilously near the cat flap) and it contained wood pellets. Whilst wood is a great substrate in many ways it can be quite unpleasant for some cats. Jake was probably retaining urine, which may have resulted in the passing of larger quantities than normal when he sprayed.

So why was Jake spraying indoors? As you now know, the

act of urine spraying is a perfectly normal feline behaviour used in areas of shared territory where the sprayer feels a sense of threat or conflict. It provides an important source of the cat's own smell and is probably extremely reassuring. There should in theory be no reason to spray urine indoors – after all, a cat's home is his den, a haven of security. Sadly, in Jake's case, this safe zone fell apart when Chris fitted the cat flap and any Tom, Dick or Sooty could enter at will and beat poor Jake into a pulp.

Can you imagine Jake in a terrible state of angst trying desperately to make sense of this? Like all cats, he has a limited number of coping strategies so spraying urine in these new-found areas of conflict seems a good course of action. So he sprays and then gets walloped by his owners. His world is rapidly falling apart as everything he thought he could rely on for comfort and security is turning against him. No wonder he is pacing and crying – I think I would have been by then.

The behaviour programme

I explained my thoughts to Laura and Chris (I tried not to make them feel too bad about the punishing bit, everyone does it) and I devised a plan that we hoped would restore some sense of safety for Jake. His confidence had been seriously shattered and we really wanted to try to restore it. Here is the plan that we put in place and followed for a couple of months.

- Laura and Chris were asked to visit their veterinary surgeon to have Jake's urine analysed to rule out any potential urinary tract problem. Whilst this was not a likely cause for the behaviour, it was important to check.
- The feral colony was a major cause of the problem and many of the neighbours were experiencing similar difficulties. A breeding colony can harbour many diseases such as Feline Leukaemia Virus (FeLV) and Feline Immunodeficiency Virus (potentially AIDS) and a bite from a carrier can mean your pet will become infected. Feral cats tend to be very territorial and correspondingly aggressive, so vet bills for bite wounds and abscesses are common. Many charities will humanely trap entire feral cats to neuter them and treat or put to sleep any that are sick. Laura put the wheels in motion to deal with the problem and had the help and support of many of her neighbours.
- Chris's handiwork had to go. The cat flap was immediately removed and the door panel replaced with a solid piece. It isn't enough to block up the cat flap since this does not remove the visual cue of the vulnerable opening.
- Jake started to come and go through the bathroom window again, and the front door or window. This meant he was restricted to excursions when his owners were home but, in the circumstances, this seemed sensible. It also represented an important message for Jake. If any cat was to come and go they had to ask his owners first. We had to make sure

Jake started to trust them again since their relationship had suffered a little with all the shouting and smacking.

- Laura and Chris were asked to use the synthetic feline facial pheromone spray according to the manufacturer's instructions to treat the areas where Jake had sprayed urine. They also used the plug-in version in the hallway.

- Jake was given an additional litter tray upstairs in the spare bedroom and the original tray was relocated away from the back door in a discreet corner. Both trays contained a fine grain substrate that was easier underfoot.

- Laura used a cleaning regime for the soiled areas of carpet in the living room and hallway to remove any residual smell of urine that would have encouraged Jake to return.

- We needed to get Jake more active and alert indoors. (Remember the importance of play therapy?) We gave him boxes to explore and both Laura and Chris played games with Jake to take his mind off all his problems and restore the relationship between owners and pet. I also asked them to make a catnip toy!

- I asked Laura and Chris to review the house to make sure there were plenty of warm hiding and resting places for Jake. He particularly liked jumping up onto high surfaces so we made an area for him on top of the wardrobe and lined it with an old sheepskin coat. This proved very popular!

- Jake was being fed a mixture of wet and dry food so we gradually removed the wet content and started to feed the dry food only. This was a good quality veterinary formulated diet (light formula as he wasn't getting out much) and I made sure that Laura measured it accurately to avoid Jake's becoming overweight. I recommended food-foraging techniques (see Chapter 11 for full details) to challenge Jake and provide him with positive messages around the home to rival his more negative urine marking.

- Water bowls were placed around the house to encourage Jake to drink more. Water intake is essential on a dry diet and several choices of water away from the food will often be attractive options.

The outcome

Laura and Chris noticed a difference in Jake within twenty-four hours of my visit. He was spending more time downstairs and seemed less agitated. Clients often report this phenomenon of an instant change but I cannot take credit for that, unfortunately. My visit enables owners to understand the problem and they instantly become less stressed themselves and more likely to offer their cat love and reassurance instead of the punishment and resentment that had become the norm. This can have a dramatic effect and Jake was obviously feeling the rush of love from his owners!

The programme worked extremely well and, after a few incidents, Jake became more active and responsive to his owners. His spraying soon stopped completely but he continued to remain indoors for prolonged periods. Maybe he felt the risks of exploring outdoors still outweighed the pleasures. The feral colony is continuing to prove a problem with numbers far greater than Laura's group initially anticipated. The local charity is, however, doing sterling work, although there is probably still a great deal to do before pet cats will feel safe in that part of the world.

Gromit and Shaun – the case of the warring Burmese

There are often occasions in multi-cat households when it is best for everyone if the individuals are separated and new homes found. In this next case, as in many others, this just wasn't an option for the owners. They were adamant that the cats would not be re-homed and they would even consider putting up with the problem rather than feel forced into parting with their loved ones. Charlotte made this plain to me when she called me about her Burmese. She had a severe long-standing problem with urine spraying indoors and she had consulted many experts for advice with no success. Her veterinary surgeon had prescribed two different potent anti-depressants but neither had had any positive impact on the behaviour whatsoever. I agreed to see her and the family but warned, despite Charlotte's protestations, that re-homing might be inevitable if she wanted to stop the problem.

Charlotte and her husband, Jeremy, shared their home with three beautiful blue Burmese, Wallace, Gromit and Shaun. Wallace and Gromit were seven-year-old brothers and Shaun was six. All three came from the same breeder and had been lively, confident and friendly kittens. When Shaun arrived he was greeted with some suspicion but any attempts by Wallace and Gromit to intimidate him or impose a few house rules fell on deaf ears. He was extremely confident and would barge past the others to get a cuddle from his owner. They lived in a quiet residential area and the cats enjoyed access to outdoors via a cat flap that their owners had fitted four years before my visit. The original manual flap had been replaced with a magnetic one when local cats were coming in and stealing the Burmese's food.

Wallace and Gromit had always enjoyed a good relationship. They were often found grooming each other or sleeping together until a few years before my visit, when Wallace started to spend long periods of time outdoors and also seemed to distance himself from Charlotte and Jeremy, only approaching them for affection on rare occasions.

Shortly after Jeremy fitted the cat flap there was an incident in the back garden between Gromit and a neighbour's cat. They were squaring up for a fight and Gromit was all fluffed up ready to go. Jeremy saw what was going on and he went into the garden to stop things from escalating. The invading cat ran away but Wallace and Shaun followed Jeremy into the garden to see what all the commotion was about. Gromit instantly turned on Wallace and Shaun and a noisy fight ensued. Within half an hour everything was back to normal but a few days later Gromit was seen spraying urine on a new television in the living room. This seemed to herald the start of the problem and Gromit was soon spraying urine on doors, windows, curtains, electrical equipment and furniture. After a while Jeremy and Charlotte became suspicious that all the marks they were finding round the house didn't come only from one cat. Sure enough they started to observe Shaun spraying too, only he wasn't just passing small amounts when he did so. He was emptying his bladder, and pools were forming on the carpet. The house had definitely started to smell. To make matters even worse Shaun and Gromit were fighting badly at least once a week. The tension in the house was palpable.

I was able to observe Shaun and Gromit (Wallace was out) whilst listening to Charlotte and Jeremy. I watched them out of the corner of my eye and I didn't like what I was seeing. There was a great deal of staring and subtle posturing between them with each vying for whatever the other had at any one time. If Shaun went to Charlotte for attention, so did Gromit. If Gromit went to play with a particular toy, Shaun took it off him. It was passive aggression at its most effective as each cat psychologically chipped away bits of the other's emotional well-being.

The facts of the case

I learned some very revealing facts.

- This had really started when Shaun was about two years old.
- Wallace had started to spend more time away from the house and become less sociable with his owners when he became a mature adult.
- All three cats were male Burmese of a similar age.
- The problem had been present for four years.
- Gromit's altercation with another cat resulted in a fracas with Wallace and Shaun.
- There were many cats in the neighbourhood.
- Shaun and Gromit preferred to stay in at night.
- There was a single litter tray available in the kitchen that Shaun used to use rather than go outside.
- Shaun was spraying large quantities of urine.
- The spraying was widespread.
- Gromit was spraying urine many times a day.

This was a huge list of relevant facts but it was a complex case and many aspects were influencing the problem. From day one this multi-cat household was a recipe for disaster. We had three male cats of a similar age co-habiting. The brothers, Wallace and Gromit, were fine until they became mature (social maturity occurs between eighteen months and four years of age), at which time there was an obvious cooling of the relationship.

Many cats vote with their paws when they become adults in a multi-cat environment and start to seek alternative 'dens' that they can call their own. Wallace started to spend increasing amounts of time away from home and I suspected that if Jeremy and Charlotte made enquiries locally they would find that Wallace had several part-time homes where he rested and had quality time alone. Like many cats in the same situation he would still return to his original home, albeit infrequently, but in such cases it is perfectly usual for the relationship with the owners to cool. Wallace would have known that access to Charlotte and Jeremy was controlled by the more assertive cats in the household and he wouldn't have taken any chances. He mustn't rock the boat, after all.

Gromit was a home-loving cat and he had no intention of leaving. Unfortunately Shaun liked his home comforts too and once he matured socially it put an extra strain on proceedings. By this time all three cats were acutely aware of the presence of other cats and feeling a strong sense of overcrowding. When Gromit became excited about the cat in the garden it wasn't surprising that he redirected his aggression onto the other two when they came to investigate. Whilst this type of spontaneous kill-or-be-killed reaction is an instinctive response to danger rather than a premeditated act, it can still lead to a deterioration of already unstable relationships. The fighting between Gromit and Shaun continued on a regular basis. If cats are evenly matched and neither backs down to the other it can result in regular bouts of nasty fighting.

We thus had a pot of stress threatening to bubble over. The combination of the sense of overcrowding and the cat flap (a possible point of invasion) became too much to bear for Gromit and he needed to express this intense emotion of conflict and insecurity, so he utilized the innate behaviour of urine spraying. He was obviously in such a state that he sprayed repeatedly every day. I've seen many Burmese behave in this way. Spraying becomes like breathing to some of them and I don't think they even realize they are doing it half the time. It becomes a tool for all sorts of stress-busting from frustration when breakfast is late to anxiety when the bedroom is redecorated. It is a real nightmare to clean up after these cats. I have to say that the all-time record in my caseload goes to a Burmilla (originally a Burmese crossed with a Chinchilla) who sprayed a total of forty-five times in one day!

Shaun, similarly affected by all these territorial issues, followed suit shortly afterwards. However, Shaun had other complications going round in his head. He found the whole concept of outdoors rather challenging, particularly if he had to perform his bodily functions at the same time. He had obviously decided that this was a seriously flawed habit so he always returned indoors, from his brief excursions into the garden, to use the single litter tray in the kitchen. As part of the continuing psychological battle between Gromit and Shaun, the former had taken to nonchalantly guarding the tray area to prevent Shaun from gaining access. This left Shaun with his legs permanently crossed so that when he sprayed urine indoors he also voided his bladder. When a large quantity of urine is passed, diagnosis of the problem becomes confusing. Is the cat spraying or urinating? I would suggest it may be a combination of both. The added complication is that cats with urinary tract problems will often stand to urinate so it's important to rule this out as a cause. Charlotte and Jeremy's vet had been very thorough and there was no question that any of the three cats had an underlying medical problem. Shaun couldn't get to the tray because Gromit was guarding it so he sprayed and peed at the same time.

I really liked Charlotte and Jeremy and my heart went out to them when I considered the possible solution. The cats had to be separated; I couldn't believe that this situation would ever resolve. Needless to say they were adamant that this would not happen and they vowed to do anything and everything to ease matters. It wasn't the damage to the home that they were stressing about. They couldn't bear the thought that their cats were unhappy. I tried to explain that their cats wouldn't *be* unhappy if they were separated but that is a hard concept for any owner to grasp. That was too much like rejection, giving up, failure, loss and all those other complex emotions that owners experience at these times.

The behaviour programme

So I devised a programme and told Charlotte and Jeremy that it was the only chance we had to keep the cats together.

- I asked them to install some synthetic facial pheromone devices that emit a calming message for cats. (In my experience these plug-in diffusers rarely solve problems when they are used in isolation but they can be extremely effective in combination with behaviour therapy.)
- They were asked to confine the cats to the kitchen and conservatory when they were out and only to allow access to other rooms under strict supervision.
- Charlotte placed two additional litter trays in the house, one in the conservatory and one in a discreet corner of the bath-room. With the best will in the world Gromit couldn't guard all three.

- A thorough cleaning regime was recommended for all the areas that were starting to smell rather unpleasant. Shaun's pee was soaking into the carpet in various places and the heady aroma was overwhelming on occasions.
- All areas that had been sprayed with urine were cleaned with hot water only and then sprayed with a mist of surgical spirit. They were then treated with synthetic feline pheromones.
- Jeremy and Charlotte went through the whole house and added a vast array of cat goodies. They provided extra beds, high resting places, scratching posts and tall cat 'climbing frames'. They also made sure that certain areas such as under beds, inside wardrobes and the bottom of the airing cupboard were all accessible for those private moments when the cats didn't want to be disturbed.
- They also started to feed the cats dry food in various locations, and water bowls were placed in every room.
- It was extremely important for Charlotte and Jeremy to

change the way they interacted with their cats. Shaun and Gromit were constantly seeking attention and their owners were so accommodating that their every desire was always fulfilled. Most of the fighting between Shaun and Gromit took place when they were competing for affection so it was important to try to give them alternative interests. Charlotte and Jeremy were told to withdraw from the relationship and be prepared to ignore or reject their advances. Contact would be only on the owners' instigation from now on.

- Charlotte was keen to get Wallace back into the fold so I recommended that, every time he returned, she should make some special time to be with him on his own for play and grooming. He always used to love being brushed but Charlotte found that he would now growl or run off if she tried to do it in front of the other cats. I asked her to take Wallace to another room if necessary so that he could have the attention without interruption.

- It was essential that the cats were given plenty of other things to do so they were provided with toys and climbing frames and challenging feeding opportunities to distract them from the more negative issues of antagonism.

- If Shaun and Gromit started to fight I suggested that they could be interrupted by a loud noise or a cushion placed between them. (It is always foolish for an owner to separate fighting cats by using a body part. I guarantee this will hurt tremendously as one or other – or both – of the cats is bound to cause injury.)

- Poor Charlotte and Jeremy were getting very distressed themselves about all the fighting and urine spraying, and this emotion could easily fuel the aggression and tension between the cats. I asked them to try very hard to relax; the cats hadn't been seriously harmed yet and that would probably remain the case.

- Medication had already been prescribed for the cats and it hadn't helped. Charlotte and Jeremy were therefore rather disillusioned about drugs and reluctant to try any alternatives. I felt it would be useful to consider something gentle so the referring vet decided upon a selection of Bach Flower Remedies (see Chapter 2) for Shaun and Gromit.
- I asked Charlotte to report twice a week with a comprehensive diary of progress. I knew this programme was our only chance of success and I wanted to keep on top of the situation.

The outcome

Charlotte was a star and she reported fully by email twice a week. There followed an emotional rollercoaster of good days and bad days but gradually things started to improve. Wallace definitely started to come indoors more and seemed to relish his private time with his owners. Shaun and Wallace were still fighting but both Jeremy and Charlotte were employing a variety of tactics to distract them. The urine spraying was improving (Shaun had chosen the litter tray in the bathroom as his favourite) but it continued on a frustratingly regular basis. Charlotte would often send an email of utter despondency cataloguing fights and tantrums, followed a couple of days later by one of elation when she reported 'no spraying found and all quiet'. It really was very difficult to see any pattern in all of this. I got the impression that we were merely offering the odd distraction but the cats' fundamental inability to cohabit remained. However, towards the end of the eight-week period things seemed to consistently improve. I still cannot really say what triggered it but the household became calm for the first time in years.

At the end of the programme I spoke to Charlotte to discuss plans for the future. This wasn't a miracle cure; there were still occasional incidents of spraying and Gromit and Shaun were still not happy together. In relative terms, however, things were much better and Charlotte felt that she wanted to persevere. The extra work for both Jeremy and Charlotte was immense but they still felt the inconvenience was worth the fact that they could (for now at least) keep all three cats together. If the owners' emotions in these cases were not a consideration it would always be preferable to separate cats like Gromit and Shaun and place them as single cats to live out their days in bliss. In reality it is impossible to ignore an owner's emotional attachment and cases often are completed with a high degree of compromise. It is pointless for me to insist that cats are re-homed. Nobody would listen to me and bad situations would continue. At least if I get involved I may be able to improve things.

CHAPTER 6

Aggression

IT IS HARD TO IMAGINE (UNLESS YOU HAVE EVER BEEN A VICTIM yourself) the extent of the injuries that an aggressive cat can inflict. We tend to think that, given their size, the resulting wounds will be superficial and inconsequential. Over the years I have developed a healthy respect for the domestic feline and I don't go for bravery awards when confronted with a potentially dangerous animal. I use body armour and show extreme caution. Basically I'm scared because I know what they can do. A penetrating bite wound from a cat usually doesn't bleed profusely as the pointed canine teeth create a puncture hole that quickly heals over. All the unpleasant bacteria from the cat's mouth are then trapped in the tissue under the skin and infection is almost inevitable. I spend every working day trying to avoid getting bitten. The scars do not prove that you are a fearless person; they merely illustrate that either you didn't read the signs of impending attack or you are a deranged masochist.

Before labelling your own cat as a nasty piece of work it is important to understand that there are many forms of aggression, not all of which are as sinister as they appear. Aggression is such an intrinsic part of the cat's survival strategy that it is almost inevitable that your own pet will exhibit a degree of aggression at some stage in its life. It is useful to look at the more common motivations for the behaviour and therefore the measures necessary to try to control the problem or stop it before someone gets seriously hurt.

Play aggression

Kittens exhibit play behaviour from the moment they open their eyes and start to move around their nest. Play is an inherent part of the animal's development and it mimics skills utilized in predatory and social behaviour. Kittens fight each other and pounce on objects as if they were prey. Enthusiastic rough-and-tumble fights between siblings are tempered if they become a little too violent; kittens soon learn to inhibit their biting in the name of 'play'.

Unfortunately when a human becomes part of the sequence we are usually not quite so good at rewarding the more measured approach. We have a nasty knack of reinforcing the highly aggressive stuff and kittens can easily develop into difficult pets if they haven't been taught to play nicely. Men are particularly fond of interacting with kittens in this way; they use their hands to grab and roll the cat around in a frenzied 'attack'. The game then becomes an exciting burst of violence as the front paws clasp the fingers and the hind legs rake repeatedly at the wrist. Human blood is often shed at this point but men seem impervious to any pain inflicted by a tiny animal and the sequence continues without interruption. An exhausted kitten eventually ends the game having learned nothing whatsoever about the boundaries of acceptable behaviour. This is how feline hooligans are created. The kitten grows up into a strong adult with formidable teeth and claws and all human hands and feet soon become potential targets for kicking, scratching and biting. The owners are repeatedly ambushed in the dead of the night on the way to the bathroom by a marauding moggy intent on violence and mayhem. The cat is then labelled as a 'psycho' and a cat behaviour counsellor is called in to pick up the pieces or the animal is re-homed with the caution 'only suitable for a household without children'. These cats are not fundamentally aggressive but they need a fair amount of consistent rehabilitation before they can be trusted to demonstrate a more acceptable response to their owners' flesh.

Play aggression is easily prevented by ensuring that human body parts never form part of a game with your pet. There are numerous toys on the market, many of which are attached to rods or sticks to allow easy manipulation. Cats enjoy the movement and find these toys very exciting, and any hands or feet are kept well away from the action. Hands are then associated with gentle stroking, holding and feeding rather than anything more sinister.

Assertive or learned aggression

There is another form of aggression that we inadvertently

teach our cats but it needs a particular combination of owner and cat personality for it to rear its ugly head. Some cats are extremely confident and assertive and, given the right circumstances and an extremely gentle and compliant owner, they can become controlling and manipulative using aggression and the threat of violence.

These cats are often restricted in their activity outside and spend the majority of their time indoors with a caring (usually female) owner. Every approach they make is rewarded with attention, food and complete submission. This power to control the owner can be addictive for some and they soon learn that the most immediate response and total compliance can be achieved through posture and aggression. I have visited many homes where one person (usually female and the most devoted carer) is herded around the house and victimized by the family pet. Believe me, that is only funny if it isn't happening to you.

Cats that exhibit assertive aggression have to understand the chain of command within the household. The person paying the mortgage really shouldn't be at the beck and call of a small animal that performs no significant role in providing for the family and doesn't even pay rent. The target of the aggression needs to change the way she (or he occasionally) reacts to the cat to illustrate that the behaviour is unacceptable, unrewarded and therefore pointless. Aggressive posturing is only relevant if the victim is taking any notice; otherwise the cat just looks stupid. A period of healthy neglect, together with the protection of stout footwear and clothing, usually does the trick. If this strong signal of 'non-reward' is consistent and the cat is offered opportunities to indulge in other more natural and acceptable pastimes it is entirely possible that the problem will resolve for good.

Fear aggression

Aggression can be used both offensively and defensively. Fear-based aggression is purely utilized as a survival strategy in circumstances where the cat feels vulnerable and in danger. Many cats deprived of early socialization with humans will remain fearful in their presence and unwanted advances and attention can result in the use of aggression as a deterrent. If the threat of aggression does not result in a withdrawal then unfortunately teeth and claws will be employed to enable the cat to escape. In these situations the humans can usually control the aggression; leave the cat alone and it won't do any harm. Unfortunately if the cat is injured or needs to be handled for its own safety then it is often necessary to take precautions against a potential attack.

The best resolution for fear aggression is to remove the need to feel fear in the first place. This requires patience and gradual exposure to the source of that fear so that the cat learns that people are actually not as dangerous as he first thought. Unsocialized cats are at their happiest when they are co-habiting with you rather than forming one half of a major relationship. If you ignore him by offering no eye contact, verbal communication or direct approaches then he will soon feel less threatened and therefore be less likely to exhibit aggression.

Redirected aggression

There are occasions when it is possible to be in the wrong place at the wrong time and this will certainly have been the case if you have been the victim of redirected aggression. This behaviour appears, at first sight, to be an unprovoked attack from a previously amiable and gentle pet. It is a frightening and hugely disappointing scenario as many owners feel their cats suddenly hate them with an intense ferocity for no apparent reason. The aggression can sometimes continue for a significant period.

Such attacks are never unprovoked, but they are certainly not premeditated or planned in any way. A typical case would develop something like this. A normally placid cat called Harry is sitting in front of the patio doors staring into the garden, his tail swishing slightly from side to side. In the distance can be seen next-door's cat, a viciously territorial creature, harbouring bad thoughts about the ongoing dispute with his neighbour. He approaches the house, having spied the said adversary through the patio doors, to pull faces at him and generally get on his nerves. Harry experiences an intense emotional overload as the cat approaches the glass and his body prepares him for a fight as the adrenalin pumps. But the cat is the other side of the glass and Harry is like a coiled spring waiting to kill him at the first opportunity. He starts to mumble and growl under his breath and his owner rises from an armchair to comfort him and reassure him about his safety given the thickness of the glass between him and his arch enemy. As the owner reaches out and touches him something extraordinary happens. The stimulus of the touch on his coat has triggered the attack that was meant for Harry's feline adversary. In an explosion of teeth and claws the poor owner's flesh is radically redesigned as Harry fights for his life without even stopping to see that he is actually attacking his beloved owner. Oops! Harry is probably mortified afterwards but certainly confused, and the poor victim is in Accident & Emergency.

Time heals these cases and owners soon learn not to touch

their cats when they are fighting imaginary battles with others outside. Often cats will burst through the cat flap after a potential altercation and then proceed to take it out on their owners, but the signs are always there to stay well out of reach until the adrenalin has stopped pumping and the cat has calmed down. Occasionally the cat will get locked into this intense emotional response of aggression and the same behaviour is triggered every time the owner enters the room. This scenario needs careful assistance from a professional pet behaviour counsellor to ensure that the cat returns to a pattern of more positive associations with its owner. The cat probably leads an existence that is devoid of appropriate stimulation and a lifestyle change may well be on the cards.

Pain-related aggression

Aggression can also be motivated by pain and disease and this is another example of the importance of veterinary investigation into all unusual behaviour before attempting a DIY solution. If your cat has been a gentle soul previously then it is entirely possible that the aggression is exhibited as a result of discomfort. Aggression is used as a warning in these circumstances because the cat feels vulnerable and your touch may well be causing the pain to intensify. If the pain is controlled or removed then the aggression will undoubtedly go with it.

Idiopathic aggression

Idiopathic refers to conditions where the cause is unknown. These are dangerous cases since the cats will potentially show unprovoked aggression at any time. If a trigger cannot be identified, or the aggression is accompanied by bizarre behaviour before or after an attack, then it is possible that it has a physical rather than a behavioural cause. These are not situations to tackle alone; cats exhibiting idiopathic aggression can be extremely dangerous and measures should be taken quickly to ensure everyone's safety. A veterinary examination is essential.

Predatory 'aggression'

I have put this in inverted commas because, of all the different motivations for aggressive behaviour in our pet cats, their response to small prey is probably the most natural. It is a testimony to their incredible hunting skills, but if you are having a problem with corpses in your kitchen, and really cannot bear it, there are a few suggestions that you may find useful.

- It may help to confine your cat indoors at those times when his or her hunting trophies have been at their most prolific, such as during dawn and dusk and at night. Many cat charities recommend that cats are kept in at night in urban areas so this is a fairly sensible routine to adopt anyway. If your cat is not used to staying indoors then build up the length of confinement gradually; for example, one hour the first night increasing to two hours by the third night and so on. When your cat is indoors it is important to offer plenty of play and games to compensate, as well as warm beds and quiet resting places. Playing with furry toys and feathers on sticks will not make your cat a more proficient hunter.

- Try adding two small bells to your cat's collar. These may knock together and alert any potential prey to the danger. Many cats get wise to this very quickly and hold their necks so still a large cowbell wouldn't sound if it were dangling there.
- Ultrasonic collars are available that emit a high-pitched sound to warn prey but these also are not 100 per cent effective.
- Even if you are a bird lover it is probably best not to encourage birds into your garden with feeders and bird tables. If you feel compelled to do so then it is advisable to make the stand of any table as tall as possible.
- If a small rodent is running around your kitchen then ensure you attempt its capture using an old oven or gardening glove. They bite.
- Worm your cat regularly. Prey animals are often hosts for certain parasites and hunting can lead to a revolting worm burden that could become evident from either end of your cat.
- Do not feed your cat more food to stop him hunting. The desire to hunt is not triggered by hunger.
- Do not punish your cat for bringing in prey, dead or alive. He is merely doing what comes naturally and choosing to bring food back to the den.
- Watch your outside fishponds and protect them with netting. Your cat may be into fishing too.
- Don't think ill of your cat if he appears to play with his prey by tossing it up into the air. This is not a sadistic streak but merely a survival strategy. Most small furry things bite back and your cat is handling the creature in the best way to avoid injury.

❀   ❀   ❀

Here are a couple of cases that illustrate both play and assertive aggression rather well. If you can relate to these stories I hope they will give you some ideas for tackling your own situation. I personally feel, where aggression is concerned, that any owner is perfectly justified in seeking professional help at the earliest opportunity.

Monster – the case of the aptly named cat

Have you ever noticed how some cats are very aptly named? Tiger will often look like a little tiger and Fatso is destined for a lifetime of obesity. So why oh why would you call a cat Monster? Don't you think this is asking for trouble?

Monster was a six-month-old tabby cat. Jacky was a mature student who lived alone. She had decided that she wanted the company of her very own cat for those long evenings hunched over her text books and essays. She had always lived with cats as a child but this would be her first for some years. She wasn't quite sure where she should look for such things but she went to her local pet shop and there she found a litter of three rather pot-bellied kittens. Two were black and white and playing eagerly together and one was tabby and sitting with his back to the others. She felt sorry for this little one so she took him home. He spent his first few days hiding behind the sofa but Jacky was undeterred. When he did burst forth from his hiding place it was as a spitting and hissing fur ball that promptly chewed and scratched Jacky to pieces. 'You little monster' (or words to that effect) confirmed his name and Monster was finally settled in his new home.

Poor Jacky hadn't really bargained for Monster's rather extreme personality. Did he love his new owner or hate her? She found his behaviour very confusing. He would constantly pounce on her feet and chew them; she would often walk round her flat looking as if she was wearing one huge furry slipper. When Jacky tried to stroke or cuddle Monster he would launch an attack with teeth and claws then run around the room with big round eyes and a tail like a lavatory brush. She had great plans for disciplining her new kitten that included *not* jumping onto the table and *not* chewing the house plants and *not* shredding the wallpaper and the arm of the sofa. Fat chance! He jumped on the table, chewed the house plants, shredded the wallpaper and redesigned the furniture. Whenever she shouted at him he didn't pay a blind bit of notice. At one point his fur fluffed up so much, enlarging him to such terrifying proportions, that Jacky locked herself in the bathroom to escape.

Monster never stopped – if he had been a child Jacky would

have been suspicious of a diagnosis of Attention Deficit Hyperactivity Disorder. She even checked the labels on his food for any food additives or E numbers. All she wanted was to cuddle him and love him but he didn't stand still long enough. She was pretty disillusioned by the time she called me and I visited her neat little flat.

Monster and Jacky shared a one-bedroom flat in East London. The rooms were fairly large and there was a staircase down to the door on the ground floor giving Monster plenty of fun and exercise. On closer inspection it was clear to see that her home lay in tatters. The only pristine thing present in the whole flat was the recently purchased scratching post – completely ignored by Monster. The wallpaper had long parallel scratch marks on it, the sofa and chair were shredded and soil was spread about on the floor around the chewed house plants to show where Monster had been digging. He was fascinated by me but I completely ignored him (for a very definite reason) and he soon focused his attention on my magic bag. That kept him occupied for a while as I spoke to his owner about the problem.

The facts of the case

The revealing facts were as follows.

- Monster was purchased from a pet shop.
- He was kept exclusively indoors.
- He pounced on feet and hands.
- He resisted approaches from Jacky by using aggression.
- He was very active and confident.
- Jacky was home most of the day and very caring/compliant/loving.

Jacky was a very attentive owner who constantly acknowledged Monster's every approach and movement about the flat. Many cats thrive in an environment where they are the constant focus but there are always those who find this extremely stressful or annoying. Monster was a very active and intelligent kitten and he was exploring his surroundings and utilizing anything (or anyone) that enabled him to develop natural patterns of behaviour. Since Jacky was the only moving object in the flat it was obvious that she would be targeted for all sorts of activity that mimicked predatory or socially agonistic behaviour. Feet and hands are just the right size to grab hold of and give a good kicking, just like play fighting with a sibling or killing a large prey animal. Monster wanted to investigate the pot plants, the high table and the textures of the furniture and wallpaper and no amount of noises made by Jacky would appear remotely relevant. Monster was merely filling his day with as much activity as possible and getting rid of Jacky whenever she threatened to interrupt his games.

Sadly I got the impression that this was never going to be the sort of relationship that Jacky really wanted. When a kitten is obtained from a pet shop it is unlikely that the background or upbringing of the litter will be known. You will never get the opportunity to visit the breeder and see how the kittens spent those important first weeks of their life. As a result what you get is a bit of a lottery. Monster was a confident kitten who didn't need the close comfort and tactile affection that so many crave and thoroughly enjoy. If Jacky and Monster were to remain together then Jacky would have to change the way she behaved towards her pet. She would also have to drastically rethink what she actually expected from the relationship. We talked long and hard about re-homing Monster but Jacky was adamant that, whatever it took, she would form a bond of sorts with this rather challenging kitten.

The behaviour programme

The programme included many of the elements common to most of my therapy plans.

- Monster's food was changed to a veterinary formulated growth recipe dry diet. Many wet foods are highly palatable and nutritious but it is hard to play exciting games with a tin of salmon in jelly! Making a cat work harder for its food is incredibly stimulating and it mimics the natural acquisition of food far more accurately than two meals a day provided in one predictable location. Don't forget that the devil makes work for idle paws and I didn't want Monster to have too much spare time. I recommended to Jacky that she placed his food in small amounts throughout the flat in challenging receptacles (boxes, cardboard tubes, egg boxes, etc.) to require some intricate problem-solving when Monster was hungry.
- I also asked her to place more water containers around the flat away from the food.
- Jacky had to start to interact differently with Monster, in a way that was more sympathetic to feline social communication. I asked her to only make direct eye contact with half-closed eyes (direct stares are challenging in the feline world). Jacky should only touch Monster when he made social advances to her and even then it should be brief. This may seem harsh but I knew that Monster would find it far more appealing.
- It was still important that Monster should be used to handling so I asked her to continue to lift and restrain him for short periods. This required a knowledge and

understanding of body language and I spent some time explaining to Jacky which moments would be appropriate and which were best avoided. After all, even Monster had the occasional mellow moment.

- Jacky was asked to play with Monster but all the games needed to be remote. I suggested the use of fishing-rod toys (with very long sticks attached so that Jacky was well away from the action). These could be agitated for Monster to chase and use up some of that apparently inexhaustible energy.

- Cardboard boxes and large paper bags were scattered randomly around the flat for Monster to investigate and destroy (if necessary).

- Some cats like Monster can become quite aggressive on catnip but, luckily, Monster certainly didn't get any worse when he inhaled the intoxicating herb. I therefore recommended a home-made catnip pouch that was just the right size to kick and bite to take his mind off his owner's feet.

- Jacky also had to wear stout boots around the flat for a while since these never taste quite the same as soft flesh.

- The scratching post that she had bought for Monster was rather short and she had been active in her encouragement for him to use it. This often has quite the reverse effect and guarantees that the cat will not set foot anywhere near it ever again. I suggested that Jacky move the post to another location and mount it on top of a sturdy wooden box to raise the height. She then covered the box with bits of carpet tile and rubbed the whole thing with dry catnip. I encouraged her to play fishing-rod games around the base of the post to associate it with positive and fun activity. I also asked her to purchase another much taller one for the opposite end of the living room near the radiator.

- The relatively small flat would be Monster's world for the foreseeable future so we had to utilize the space as efficiently as possible. Jacky purchased two hard-wearing carpet runners for Monster to climb and attached them vertically to a wall in her living room using double-sided carpet tape and wooden batons at the top and bottom. A shelf was then positioned adjacent to the top of the carpet on the wall, with a series of more shelves like a staircase beneath it, so that Monster could get off at the top and use the shelves as steps down or useful resting areas. Each shelf was equipped with a furry toy that Monster could push to the ground just for the fun of it.

- Monster still needed quiet time on his own for rest, despite Jacky's belief to the contrary, so we created several private warm areas. I warned Jacky not to disturb Monster whilst he was there.

- We reviewed all the house plants in her flat and removed any that could potentially be dangerous. We then added catmint plants and grasses in containers as a useful source of vegetation.

- The existing house plants were protected with cardboard circles cut round the base of the plant to cover the soil and remove the temptation for Monster to dig.

The outcome

Jacky attacked the plan with great enthusiasm and within a few weeks she started to see the fruits of her labour. Almost instantly Jacky had seen a difference in Monster when she started to pay him less attention. He certainly was biting her less and he was concentrating on inanimate objects and toys. He loved looking for his food and he used to enjoy hooking the biscuits out of the egg boxes and toilet roll tubes. Jacky was particularly delighted when he started to scratch on the redesigned scratching post, and I recommended that she used a low-tack double-sided adhesive tape on the wallpaper and sofa to prevent further damage there.

So ultimately Jacky had an active and very playful young cat who was prone to chewing on feet and hands if he was given the opportunity. Unfortunately she had to do quite a bit of extra stuff to entertain him and prevent any 'attacks' and she soon realized how chore-based Monster's ownership had become. This is often the case with certain difficult individuals. Monster would have been just fine in an environment where he had access to outdoors; he could have as many adrenalin rushes as he liked in the big wide world. Placing a cat like Monster in an indoor situation puts an enormous responsibility on the owner to entertain him. Some feel that the price to keep a square cat in a round house is too high. Jacky didn't and, as far as I know, Monster is now mature and still very much Jacky's cat.

Woody – the case of the cat that bullied his owner

Woody was a rather handsome two-year-old black moggy who lived with Gary and Tracey and their two teenage children, David and Rebecca, in a house in East London. Tracey had got to the end of her tether with Woody because he just wouldn't stop attacking her. Gary, David and Rebecca greeted my visit with hysterical delight and the whole consultation turned into some sort of floor show rather than a sober appraisal of problem behaviour. Sometimes it works out that way and I find it important to 'go with the flow'. Tracey was obviously incredibly embarrassed about the fact that Woody targeted her exclusively with his unsociable behaviour and the whole concept of calling in a cat shrink could only really be approached with her tongue very firmly in her cheek. I can understand that.

Woody had been a birthday present for Tracey (how ironic) just over two years ago. He was the product of a liaison between a small female tabby cat in rural Essex and a huge black panther-like creature that stalked the local area. The little tabby had litters with exhausting regularity and all her charges were born and reared in a hay barn. The owner tried to handle the kittens as best she could but they were rarely seen until they were four or five weeks old. Gary chose Woody because he appeared to be fairly confident and inquisitive and he felt he would make a great pet. He was also jet black like Tracey's family cat when she was a child and he felt this would be a big bonus.

Tracey was absolutely delighted with her birthday kitten. Gary and the children were out all day but Tracey worked part-time so she was able to devote a great deal of attention to the new addition. However, it did take her a while to get used to his little idiosyncrasies. When she fed him he metamorphosed into a wild beast, crouched over his bowl growling and hissing as he gobbled huge chunks of food. Every approach from her would be met with a hiss or a growl. She excused this unpleasant hostility with the thought that he was a farm cat and needed to understand that humans meant him no harm. So Tracey continued to love, nurture and protect him with absolutely no positive feedback from Woody whatsoever. She fed him tasty food three or four times a day, she allowed him out when she was around to keep an eye on him, and generally remained at his beck and call. Woody continued to show aggression and there was no doubt about the identity of his victim. Tracey copped it, every time. You have to admire the strength of her determination, but surely she must have realized that her approach just wasn't working?

Meanwhile the family were sniggering at her misfortune. When I spoke to Rebecca and David about their relationship with Woody they didn't seem entirely clear about what I was getting at. Relationship? Both teenagers were into television and computer games and the majority of their leisure time involved minimal movement with eyes fixed on a screen. If the cat had entered the room and performed an elaborate song and dance I don't think they would have noticed. Woody was there, he attacked their mother and he ate smelly food. That was about as much as they could tell me about their relationship with the family pet.

Gary was equally dismissive. He respected the fact that his wife loved the cat but pet care, as far as he was concerned, wasn't really his domain. I asked him what contact he had with the cat in a typical day and all he could think of was pushing Woody off the sofa when he got home from work so that he could watch the television. Probably the most annoying fact that came to light during my visit (as far as Tracey was concerned) was that Woody appeared to have a healthy regard and genuine affection for Gary. I felt a pattern was emerging.

There was no question that Woody was getting progressively worse. Several times now Tracey had been cornered at the bottom of the stairs and been unable to move because of Woody's menacing stare. When she had attempted to pass him she had paid the price with lacerations to her ankles. She now felt she was a prisoner in her own home and she was totally at the command of her beloved but rather sinister pet. Tracey was frightened and she seemed powerless to change the situation. Gary and the children were perfectly capable of removing Woody from his threatening position and thought nothing of bartering for her release. 'What's it worth, Mum?' was a familiar phrase in that household!

The facts of the case

Every truly significant fact is always highlighted in my notes when I consult. The important ones in this case were as follows.

- Woody had little experience of humans when he was very young.
- Tracey focused totally on him and was always trying to touch him and stroke him.
- Woody appeared more relaxed when Tracey wasn't in the house.
- Woody seemed to like spending time with Gary.
- Tracey restricted Woody's access outdoors despite his apparent desire to go out as often as possible.

As I have said, a cat's personality is formed from both genetic and environmental influences. The genetic element will dictate whether a cat is reactive or quiet and the early environment will provide the blueprint for future experiences. The most sensitive developmental period is between two and seven weeks of age and during this period a kitten can quickly form social bonds with other species. If an individual is deprived of this opportunity it can lead to an antisocial or fearful response to humans in the future, no matter how kind or gentle they are. Woody was deprived of this opportunity and he will remain wary and suspicious of human interaction. His fundamental character is strong and confident so he is more likely to fight than to withdraw or show fear. If a cat does not understand human/cat interaction he can only judge a human's behaviour by feline standards. A typical human approach of eye contact, outstretched hands and constant focus would, in cat terms, be a threatening and unpleasant gesture. People who largely ignore cats and co-exist rather than trying to make friends are non-threatening and highly attractive in comparison. This was why Woody was perfectly happy in Gary's presence.

Women tend to be more attentive and solicitous towards their pets and most will make eye and verbal contact almost continuously in the presence of the family cat. Despite the fact that most cats would thoroughly enjoy this, Woody found Tracey's attention threatening and irritating and, as a result, had a poor relationship with her. He soon learned that aggression would stop the approaches, so this became his strategy whenever she appeared. It was no coincidence that Woody relaxed when Tracey wasn't around. No other member of the family would pay him any heed and he could go about his business with the knowledge that he was safe from Tracey's prodding and poking. He wasn't too happy either about being confined. Woody loved to be outside and actively defending his territory, hunting and doing all the things he was put on this earth to do. Confinement frustrated him even further and made him even angrier.

These problems can dramatically resolve if the recipient of the attacks changes the way she interacts with her cat (sorry to use the feminine but this is a woman thing). Tracey needed to adopt an air of healthy neglect and rejection in Woody's presence by mimicking the behaviour of the other human members of the household. Tracey hated to admit it but Gary and the kids had got it completely right where Woody was concerned.

The behaviour programme

So, with a degree of mirth and merriment, we talked about the forthcoming programme of behaviour therapy.

- Tracey should start to interact differently with Woody. She should regularly ignore him. This would include no eye contact, no watching and no verbal communication. It was important she understood that Woody would not feel rejected as a result and the effect would probably be extremely positive by making him more comfortable and content in her company.
- Tracey would change the feeding regime to a food that could be left down for Woody to graze to avoid the need to constantly offer meals throughout the day.
- If Woody made a friendly approach towards Tracey she should acknowledge it with a brief stroke on the head only and a single short verbal comment. If he did actually approach in that way it was certainly all he expected or wanted.
- Tracey should consider fitting an electronically controlled exclusive-entry cat flap to allow Woody to have access to outdoors when he wanted. The area was a quiet residential estate and the risk from traffic and other dangers did not seem to justify confinement against his will. (I never force owners to do this because I cannot guarantee any pet's safety but, after some discussion, Tracey agreed that this would be the kindest thing for Woody.)
- Tracey had started walking around the house with some trepidation as she expected attacks from Woody. It was therefore essential to protect her in some way so that she could walk around in a more confident manner. It was agreed that she would wear stout leather boots for the

period of the programme and, should Woody block her path, she should slowly but confidently walk right past him in the knowledge that his attack would not cause injury. It is great in theory to ignore hissing and threatening behaviour but very difficult to carry out. I had every confidence in Tracey's determination to do the right thing.

The outcome

Gary particularly liked the programme because the key phrase throughout was 'Watch Gary. Whatever he's doing just copy him!' Tracey kept in touch and followed the programme to the letter. The transformation in her pet was nothing short of miraculous. He loved his new-found freedom outdoors and, despite some initial confusion, adjusted well to his owner's novel indifference to him. There were a few incidents in the narrow hallway but, Tracey assured me, she remained firm and ploughed slowly past his flailing paws. The family reported that he soon became more relaxed and they hardly ever heard him hiss or growl again. Sadly Tracey never achieved the close cuddly relationship that she wanted. She was fairly philosophical about the whole thing and happy to continue to co-habit with a pet that she largely had to ignore.

CHAPTER 7

Inter-Cat Aggression and Multi-Cat Household Problems

OVER A THIRD OF ALL CAT-OWNING HOUSEHOLDS IN THE UK share their home with more than one cat and the numbers are increasing all the time. The benefits of a multi-cat household for the owner are obvious. The wide variation in cat personality means that you can have a feline to match your every mood. There will always be at least one to cuddle up to and any family squabbles can be avoided by providing one cat each. Many of us feel terribly guilty when we disappear to work and leave our lonely pet staring disconsolately from the window whilst waving a tragic 'goodbye' with his paw (or so we imagine). This gut-wrenching sense of rejection and betrayal can be instantly assuaged by acquiring several cats to provide company for each other. If some of those cats happen to be 'rescued' then that merely adds to our feeling of general all-round goodness. Do you see how easy it is to become a multi-cat household without really trying? I fell victim myself and lived in a seven-cat household for some years.

Once we are surrounded by the patter of a score of tiny feet we tend to presume that they are getting as much out of the situation as we are. Sadly, that is not necessarily the case. Once again we are judging another species by our own standards of sociability and relationship needs. It's important to remember that, despite often appearing so, cats are not small people in fur coats.

One of the most endearing features of the domestic feline is undoubtedly its incredible adaptability. Many modern human lifestyles are less than ideal for cats but they still manage to create a tolerable existence for themselves. However, just because cats will adapt doesn't necessarily mean they should. The whole principle of pet ownership has to be a two-way pleasure and if the cat is prepared to make sacrifices to live with us then surely we should do likewise? I have always believed that we all take on a huge responsibility when we decide to share our homes with a cat (or several). Sadly they don't come with a manual so, to a certain extent, we have to make up the rules as we go along. Without a blueprint for their wants and needs we can only judge their requirements in human terms. The ultimate sign of love for our pets has to be a respect for the species and a desire to accrue knowledge to make their lives as pleasant as possible. With that in mind it's probably worth delving a little deeper into a social structure that is really a world apart from our own.

The social life of the cat

The cat is a complicated animal capable of forming social bonds with its own species but naturally it walks alone. Cats are not co-operative or pack hunters and their need for social contact in the wild environment is often limited to mating, rearing young and fighting to defend their territory. A contradiction arises, however, when you examine feral colonies. Much of a free-living feral cat's behaviour is dedicated to its relationship with other individuals. Cats are capable of existing in an incredibly flexible and variable group situation. Social living in a natural environment provides both a defence against predators and reproductive access to other members of the species. It also allows the individual to exploit food resources that may only be accessible to groups. These and many other strategies for survival are as important to the cat as to any other species and forming a group structure in order to exploit them to the full makes perfect sense.

When cats live with us there are several significant differences from these naturally occurring groups. We tend to neuter them and we make the decision about who lives with whom. We then place them within a territory containing, potentially, many other individuals that would not necessarily have chosen to congregate in such close proximity. If my current four cats, Bakewell, Lucy, Annie and Bink, lived in a built-up area rather than rural isolation they probably would not be as relaxed and sociable as they are. You may have the most compatible group of cats but be unfortunate enough to live in an area densely populated with other cats that upset the status quo. Despite all this, some multi-cat households work but most only get by with a degree of mutual tolerance between the individuals.

I have been challenging owners to review their ideas about

multi-cat households for some time now, speaking at various seminars and writing articles saying that the behaviour cases I saw were merely the tip of the iceberg when it came to problems in multi-cat households. In 2000, with the help of *All About Cats* magazine, we conducted the Feline Felons survey, mentioned in previous chapters, encouraging readers to complete a questionnaire if they felt they were experiencing problems with their cats. A total of 267 owners took part, owning between them 784 cats (with one lady sharing herself and her home with 30 furry friends). The questionnaire asked several questions about the household in general, nutrition and lifestyle and then asked owners to list the various problems that they were experiencing.

The majority of owners (73 per cent) had more than one cat and provided a great deal of useful information, not least relating to the potential pitfalls of multi-cat households. The most common behaviour reported by two-thirds of the multi-cat owners related to territorial aggression and fighting. Sixteen per cent of the cats that fought outside were also aggressive to other cats in the household. If you are currently giving a huge sigh of relief because your cats are not fighting you may want to remember this: owners are often blissfully unaware of friction between their pets because they judge relationships to be good if there is no fighting. Cats know there are far better campaign strategies for battle! Inter-cat aggression can be passive, subtle and devious but the victim can become extremely distressed by the relentless conflict. Cats have limited ways to express their emotional states so sometimes it is useful to diagnose disharmony by default if behavioural problems can be identified within the household.

Social maturity

It is sometimes hard to understand that we are not necessarily doing the best thing for our cat when we consider acquiring a companion for him. Kittens and juveniles appear extremely social and their lives often revolve around interaction with cats of a similar age. Contact at this time is extremely important for a cat's future as it enables social skills to be developed, but it's worth noting that those skills predominantly relate to hunting and fighting.

If you have owned your cats from kittens you may have noticed a gradual change in their relationship over the years. Are they still as friendly now they have grown up? Problems often arise when cats reach social maturity between the ages of eighteen months and four years. At this time the fundamental personality of the individual will dictate its response to the presence of other adult cats and the atmosphere within the household may well change. It doesn't seem to matter whether they are sisters, mothers or brothers; all relationships appear to run the risk of becoming decidedly cold at this time. Even if you haven't noticed anything particularly untoward, I would bet that your cats have adopted one of the following strategies:

- Agreed to disagree and co-habited whilst avoiding the others at all costs.
- Been oblivious of the presence of the other cats (lucky you if this is the case).
- Decided to love everyone and remain a permanent kitten and never reach social maturity (it's so much easier).

In my experience (do bear in mind I spend my whole working life with the more dysfunctional cat families so that my experience is weighted towards them) many cats gravitate towards the following options:

- Despise the other cats and embark on a battle of psychological threat and intimidation to destroy the enemy.
- Despise the other cats but be too scared to show it (and probably eventually develop a stress-related illness).
- Be terrified of the other cats, have the word 'victim' stencilled on your forehead and get beaten up with monotonous regularity.
- Move out.

Are your cats at war?

I think you are probably getting the idea by now that I believe multi-cat households rarely work. This may seem a rather dramatic and sweeping indictment of feline co-habitation and it's important to understand that the problem is one of degree. A group that is mutually tolerant can be considered as acceptable whilst a group at war is clearly not. The matter is further complicated by the fact that cats 'at war', as previously mentioned, are not necessarily fighting with teeth and claws at every opportunity. It is not a good survival strategy to fight whenever there is conflict. Cats are armed with such formidable weaponry that every altercation could potentially be their last. Therefore they often rely on passive and covert tactics designed to disarm the enemy emotionally. For those brave enough to delve into the psyche of their cats, here are a few indications that all is not right in a multi-cat household.

- Individuals keep away from each other within the house and do not sleep or rest together.
- Members of the group respond to the presence of another by leaving the room.
- The cats will not play in the presence of another.
- One member appears to withdraw from human contact in the presence of another.
- One cat will move another away from a favourite resting place just by staring.
- One member of the group will sit in passageways, by the cat flap or on the staircase preventing other cats from moving freely around the house.
- There is active fighting in narrow corridors or at mealtimes.
- There is a fight when one of the cats returns from a visit to the vet.
- There is evidence of excessive scratching in certain areas in the house.

- At least one member of the group is obese.
- There is a history of inappropriate urination (soiling) in the home.
- There is a history of urine spraying indoors.
- One member of the group over-grooms and gets bald patches.
- One member of the group suffers from recurrent cystitis.

None of these points is necessarily diagnostic of a major problem but the presence of any of them would make me deeply suspicious that there were relationship issues within the household. Even in those groups that are working relatively well there will be social signalling taking place that could possibly indicate to a confused owner which cat is in control of the group. In my opinion there is no definitive social structure or pack hierarchy in cat groups. Owners usually feel that there is a 'top cat'; often the eldest male or female. These cats tend to spend less time with the others and don't appear to compete because there is a mutual understanding that they get what they want when they want it. Competition between other individuals will occur but the outcome will always depend on how important the particular resource is to each party. It is sometimes extremely difficult to judge which cat actually is in charge.

Social signalling

Let's look briefly at the signalling you are likely to see and its interpretation.

- A direct stare is challenging and usually employed by more dominant individuals.

- Overt fights are far more likely between individuals if there is no difference in social status.
- High-status/ranking cats may decline to actively fight. They will look away and walk away and sit and groom, indicating the other cat has lost. Only really high-ranking cats are able to do this.
- Signalling by an assertive cat can be very subtle and passive and can include anything that elicits a withdrawal or deferential behaviour. For example, standing or sitting in a doorway and blocking access to a desired area.
- 'Top cats' control access to litter trays, stay in them longer and use them first.
- Mounting behaviour can be employed as an assertive territorial gesture.
- Hisses are used to avoid frank aggression.
- Growls are agonistic, used both offensively and defensively.

Sociable behaviour

Before you get in a complete panic you may like to have a chance to identify the more positive aspects of your cats' social life. If you find that their behaviour includes any of the following then maybe things are not all bad. These are considered to be signs of friendliness and sociability between two cats.

- Chirrups when greeting social individuals.
- Sleeping together.
- Grooming each other.
- Rubbing against each other to exchange scent.
- Friendly greeting after a prolonged absence.
- Play behaviour.

If you have a harmonious multi-cat household you are a fortunate person; there is nothing more pleasant than co-habiting with several feline friends. You are in the enviable position of being able to read this chapter without concerning yourself with its relevance to you. Your cats obviously have compatible personalities and an environment that enables them all to live in an atmosphere of genuine sociability or, at the very least, mutual tolerance. However, conflict can arise at any time within a group of cats and this can lead to stress and the development of behavioural problems. Such conflict can be triggered by factors totally outside your control (new cat on the block, for example) so it is probably unwise to be too complacent. Problems can often occur where owners will not be fully aware of the cause, or haven't realized the extent of the tension between the individual members of the group.

All is not doom and gloom if you should find yourself in this unenviable position. There are lots of steps you can take that may well provide the solution to the problem. Here are a few of the more common issues you may be facing, with some general tips that may help to restore harmony.

Inter-cat aggression

Actual fighting occurs in multi-cat households at times when the cats feel they have no alternative. It tends to be between equally matched individuals but overt aggression can also occur when assertive cats persistently target those too timid to retaliate. Aggression within a household is far more likely to be passive and covert and to involve posture, staring and psychological threat. The position in which a cat chooses to rest is rarely (if ever) chosen randomly in a multi-cat household.

Sitting in a location that potentially blocks access to other areas is a great strategy to intimidate and generally stamp a cat's authority over others in the house. Other techniques include guarding litter facilities or preventing entry and exit via the cat flap. Staring at a potential victim will usually achieve an immediate withdrawal or at the very least create a sense of extreme unease.

Possible solution

Aggression is often present when the individuals within the group become competitive over certain resources that appear to be in short supply. It is therefore often helpful to review the home from a feline perspective to ensure there are sufficient resources to satisfy everyone. Things that need to be in abundant supply include

- food stations
- water sites
- beds
- private areas
- scratching posts
- litter trays
- toys
- high resting places

A good formula to apply to all resources seems to be 'one per cat plus one, in different locations'. This has always been my favoured formula for litter trays in households with soiling problems but it seems to work equally well for other provisions.

If you continue to have problems then a referral from your vet to a pet behaviour counsellor may be your next step. It's important to remember that some situations cannot be fixed and at least one of your cats may be much happier living elsewhere. This has always been one of the biggest frustrations of my job; some problems cannot be sorted because the environment and the social structure of the group just are not working. I find it extremely difficult to justify trying to keep unhappy cats together if separating them is the answer.

Inappropriate urination and defecation

Cats distressed by the close proximity of others will become highly aroused and every potential stressor appears to affect them profoundly. Cats with social problems indoors become acutely aware of the presence and potential threat of all other cats, including those in the territory outside. The prospect then of urinating or defecating outdoors can become terrifying. If there are no facilities provided in the home (there rarely are if the cats have free access outside) the fearful cat will feel compelled to utilize the next best thing and that may be something really inappropriate to us such as a duvet or sofa.

Owners who keep their cats exclusively indoors or limit their access outside will need to provide indoor litter facilities. This may still be an issue if conflict is present between the resident cats. Assertive cats often guard trays or intimidate others whilst using them and this is sufficient excuse for the victims to find alternative and safer locations. Owners may not be aware that guarding is taking place because the 'sentry point' may be some distance away. Cats are never so stupid that they would be caught performing such a heinous act!

Possible solution

It is always necessary to provide litter facilities indoors if you are experiencing a problem with soiling, even if your cats have free access outdoors. The ideal number of trays is the magic formula of 'one per cat plus one' in different discreet locations. Even the most enthusiastic bully cannot guard all of them! The trays should be as attractive as possible containing a fine grain substrate that is cleaned out daily. If some of your cats are using existing facilities then it is advisable to maintain these and put the additional trays (with the new litter, if you were using a different kind) elsewhere. A mixture of covered and open trays may be useful to cater for personal preferences. Your cat will still be drawn to the soiled areas as a result of the residual smell so a thorough cleaning regime is essential. Soiling problems are covered in greater depth in Chapter 4.

Urinary tract disease

This is a commonly diagnosed problem and many sufferers live in multi-cat households. Urinary tract disease can be related to a number of causes such as the formation of crystals in the urine, bladder stones, bacterial infection (surprisingly rare), tumours or physical abnormalities. However, the vast majority of cases have a significant stress component and this cannot be ignored.

Cats that are stressed by multi-cat living will be less active, probably overweight and retain urine. A cat is at its most vulnerable to attack when it is urinating and defecating so toilet habits can become a big issue. To a certain extent it is possible to delay a bowel movement but the bladder is only flexible to a point. Cats can hold their urine for significant periods of time but eventually something has to give. Either your cat will end up voiding the urine where he or she shouldn't in a moment of urgency, or physical problems will result. Urinary tract disease can be easy to spot or far more subtle in its presentation. Some cats will appear anxious and squat frequently whilst passing small spots of bloodstained urine. Other cats – predominantly males, due to the shape of their urethra – may block completely, requiring immediate veterinary intervention. Other signs of chronic cystitis (referred to as Feline Idiopathic Cystitis), such as inappropriate urination in the home, over-grooming and aggression, can be harder to recognize as symptoms of urinary tract disease. See Chapter 4 for more details.

Possible solution

Visit your vet as a matter of urgency. Any treatment necessary will then be carried out without your cat's suffering further. Your vet will probably recommend a wet diet and increased fluid intake. Extra litter trays (see above) and an increased number of resources within the home will also help to alleviate the stress that your cat is currently experiencing. Many vets recommend behaviour therapy to work alongside veterinary treatment for urinary tract problems and I believe this should become standard practice to get the best possible results.

Urine spraying

Urine spraying is a perfectly normal means of feline communication within a cat's territory. There should be no need to spray urine indoors since this should represent a safe haven for the residents. If there is tension between the members of the group then cats have limited ways of expressing this intense emotion and they often resort to utilizing a natural behaviour performed in areas where there is conflict. The Feline Felons survey indicated that the incidence of urine spraying indoors increased in direct proportion to the number of cats within the household. Every group of cats has that 'one too many' threshold when things start to go badly wrong.

Possible solution

This can be a difficult problem to resolve at home without the help of a behaviour counsellor. The solution depends greatly on the fundamental cause and it often takes a professional to unravel the complexities of each unique situation. Feliway (manufactured by Ceva) is a synthetic version of pheromones secreted from glands in the cat's chin and lips and can be useful in deterring your cat from spraying in a particular location. However, your cat may continue to spray elsewhere if you don't get to the root of the problem. Increased stimulation indoors and the provision of additional resources as detailed previously may help; Chapter 5 covers this problem more fully.

Obesity

It may be a bit difficult for you to equate obesity with the stress of living in a multi-cat household but the link is there. Cats become obese (just like us) by eating too much and exercising too little. A cat that finds community living a problem will often prefer to remain as still as possible and sleep more and generally stay out of trouble. Let's face it, if you don't draw attention to yourself you stand a sporting chance of not getting beaten up. You are also less likely to go outdoors because of the obvious problems of actually getting back in again and your life revolves around sleep and keeping a low profile. However, many cats need some outlet for their anxieties and comfort eating is often one of them. Dry food will often be provided 'ad lib' in a multi-cat household, in a bowl that is constantly topped up, and there is a temptation in these circumstances to eat rather more than is actually required. Even if your cat doesn't appear to eat very much it is important to remember that too much food can be a relatively small amount if your cat isn't exercising at all.

Possible solution

Most modern veterinary practices now provide clinics for those patients who need to lose weight. Obesity can lead to diabetes, heart and joint problems in later life and it represents a genuine danger to your cat's health. The practice nurse will recommend a specific diet and actual amount that you need to feed but the reduced intake alone will not lead to weight loss. Your cat needs to make a dramatic lifestyle change and that means exercise. If your cat has problems with others in the house then he may be reluctant to be active in front of them. It may be useful to follow the other advice outlined above to decrease stress and increase harmony in the home.

Over-grooming

Cats are rather reluctant to show their stress openly. Assertive cats within a household employ the psychological tactics of the worst bully and any sign of weakness in the others will be an open invitation to increase the amount of intimidation. A great deal of a cat's stress therefore is internalized and the sufferer often finds himself in need of some coping strategy just to get through the day. These techniques utilize normal behaviour such as eating, grooming or sleeping. Unfortunately

over-grooming can be quite a rewarding experience for susceptible individuals and the damage caused to the under-lying skin can be extreme. Whilst most cases of over-grooming and self-mutilation (biting and chewing causing significant trauma to the skin) result from dermatological problems or pain there is still a small percentage that appear to be totally motivated by stress. Cats that over-groom their lower abdomen are often suffering from chronic cystitis as mentioned before and licking this area in response to pain.

Possible solution
Your cat needs to be seen by your vet as soon as possible to investigate any medical cause for the over-grooming. If your vet considers this to be a behavioural problem then a referral to a specialist is essential. More information about this problem can be found in Chapter 9.

The harmonious multi-cat household

The modern pet cat appears to be suffering from some of the

same stresses as its owner including increased leisure time, lack of exercise, relationship pressures and social overcrowding. Whilst group living isn't responsible for all cat problems it is fair to say that it is often a significant factor. Some cats take great pleasure in social interaction with their own species and it's always worth remembering that no two cats (or groups of cats) are the same. What is working for you may not be working for the same number of cats in an identical house down the street.

Now we all know the potential pitfalls of the multi-cat household there are certain steps that can be taken to increase the chances of a successful and happy grouping.

- Choose compatible individuals such as litter mates, probably brother and sister. Two equal-age males may dispute the hierarchy when they mature socially.
- If you want to adopt an adult cat, choose an individual that shows a history of being sociable with other cats. Avoid those that have been given up for adoption because of indoor soiling/spraying/anxiety-related problems. Sadly, though, many cats enter rescue centres with very little information regarding their past experiences. Acquiring a second-hand adult cat always represents a degree of risk.
- Avoid extreme characters when choosing new kittens, for example extremely nervous, confident or active. They may potentially be difficult cats to live with or find others difficult to live with.
- Do not be under the impression that keeping kittens from your own cat's litter will be company for her. Once the initial rearing process is complete she will naturally be ready to say goodbye and when her offspring reach sexual maturity there could be problems.

- Keep an appropriate number of cats to suit your environment; for example, if you have a two-bedroom terraced cottage then five cats may be too many for harmony to reign. This is particularly relevant if your cats are kept exclusively indoors; your house is their entire world and everything within it becomes a potential trigger for competitive antagonism. There is no 'number of cats per square metre floor space' rule but common sense should prevail. Every cat has a concept of his own personal space and any invasion leads to tension. Cats need the ability to put a significant distance between each other.
- Try not to constantly add to a stable social group. Every household has a 'one cat too many' number and you may be pushing your luck. If you have a happy threesome for example I would strongly recommend you keep it that way.
- Avoid having too many cats in your household in densely cat-populated areas. Such situations can present a stressful sense of overcrowding that can easily cause problems even if your multi-cat household consists of only two cats.
- Avoid friendly entire male strays that appear really gentle and irresistible until they have their paws firmly under the table. They frequently enter cat-lovers' homes as passive or rather poorly specimens, but once fit and well fed they can wreak havoc on their fellow residents. Such cats, even after castration, do not do well living with others but they usually make great single pets. (See Thomas's story later in this chapter.)
- Avoid having too many highly intelligent and sensitive pedigrees in the same household. They can be extremely territorial and competitive with each other, particularly breeds such as Burmese, Bengal and Siamese.
- Remember the importance of feline-friendly resources and

start by giving your house a three-dimensional aspect. Provide plenty of high resting places to enable individuals to observe activity from a safe vantage point.

- Private places are also extremely important; every cat no matter how sociable needs 'time out' to enjoy moments of solitude. Wardrobes and cupboards are ideal and there should be plenty of choice to enable all members of the group to have their own favourite spot.

- Provide dry food for 'grazing' throughout the day or divide it into several smaller meals to avoid any sense of competition. Consider designating several areas within the home as feeding stations to avoid bullying at mealtimes. This method of feeding cannot be recommended for those cats suffering from Feline Lower Urinary Tract Disease or chronic cystitis; they need to be fed a wet diet.

- Water is an important resource to cats and several bowls throughout the home will potentially encourage your cats to visit them more frequently. Cats often find water more attractive if it is found in a different location from their food. It is advisable to encourage your cats to drink as much as possible to guard against urinary tract problems.

- Ensure there are plenty of scratching posts to protect your furniture. Cats will scratch for both claw maintenance and territorial reasons and there will be an increased need to signal to others in a multi-cat environment. These scratching posts should be located near entrances, exits, beds and feeding stations to ensure an appropriate surface is available in areas of potential competition.

- Beds in warm places are worth defending so provide enough for everyone.

- Even if your cats have access to outdoors it is still advisable to provide indoor litter facilities. If there is any bullying going on outside then your cats will always have the choice to toilet in comparative safety indoors. Providing 'one tray per cat plus one' in different locations indoors is an important formula if you already have a soiling problem in your home or a cat with a history of urinary tract disease. However, prevention is better than cure and ensuring the availability of a number of convenient litter trays will potentially avoid problems in the future.

Now you have numerous ways you can prevent problems occurring in a multi-cat household and some suggestions about how to resolve issues you may already be experiencing if things are going slightly wrong. The following cases illustrate just a few examples of the many problems that can arise, with a step-by-step guide to putting them right.

Molly, Cara, Stripey and Egbert – the case with many complicating factors

Consultations in the client's home usually take about two hours. Every now and then I visit a household where the issues are so complicated it is hard to know where to start. It is also hard to restrict myself to a two-hour period and I have often left such places completely shattered some three and a half hours later. One such consultation involved a lovely couple and their four cats whose many complicated problems involved both medical and psychological elements. Some owners get frustrated with me because I will only work on veterinary referral; they seem to think that physical and behavioural problems should be treated separately and have no link with each other. This just isn't the case and Sally and Mark's situation will clearly illustrate this.

Sally and Mark lived in Hertfordshire and had always owned cats throughout their lives. Their last elderly cat had passed away and for the first time in years they found themselves catless. Some ten years before my visit they acquired an ex-stray from their local cat shelter. She was a pretty black and white moggy and the vet guessed her age to be about two. They called her Molly and she settled in well to her new home and seemed to delight in the fact that she had found a place all to herself. Unfortunately (for Molly) Sally and Mark were softies when it came to a sob story. About a year later they heard that a local family had moved away from the area and left their male cat behind to an uncertain fate. This was totally unacceptable to Sally and Mark so they immediately found the cat, placed him in a basket and took him home. He was a fatter version of Molly with a little more white and his name was Egbert. He took his dramatic rescue very much in his stride but Molly was furious. She hissed at him and spat at him and generally danced round him with her fur standing on end but he just didn't seem to care. Sally and Mark had stumbled upon the most desirable of all cats: a laid-back equable soul, completely indifferent to the posturing of other cats and the ideal cohabiter in a multi-cat household.

As is usually the case, Molly soon ran out of steam when she saw that her efforts to remove him fell on deaf ears. She accepted an uneasy truce for a further five years until Sally and Mark fell victim to the softie syndrome again. A neighbour's elderly aunt died suddenly, leaving behind her beloved six-year-old female tortoiseshell called Cara. The aunt's relatives didn't want her and they didn't seem to want to put a great deal of time or effort into finding her a suitable home either. So, once again, Sally and Mark stepped in and whisked her away to introduce her to Molly and Egbert. Molly's rage at her arrival was surpassed by that of Cara, who was absolutely furious when she found that she was expected to share her new home with two other cats. Cara's strategy was to pretend they didn't exist and ensure that there was a distance of at least ten feet between them at all times. She found a storeroom full of boxes at the back of the house and spent her time hidden away in there or roaming outside. She took to her new owners and was extremely affectionate, but if any of the cats breached the ten-foot exclusion zone it was entirely likely that Sally and Mark would see a dramatic change in her demeanour from loving to hostile.

Undeterred by the uneasy alliance between the three cats, Sally and Mark once again succumbed to the lure of a sob story. Cat number four arrived about a year before my visit. Stripey was another female, a tabby of about eleven who had belonged to an elderly gentleman. Stripey also entered the house kicking and screaming and hated all the other cats on sight. Sally and Mark had to work very hard in the first few months to avoid bloodshed; the resident cats may have been relatively elderly but this didn't seem to dampen their determination to damage each other. Meanwhile Egbert ate and slept and appeared relatively oblivious of the battleground around him.

Sally and Mark remained fairly philosophical about the cats and felt that, as they had a large house and access to outdoors whenever they wanted, they should be able to agree to disagree and find their own space. This seemed to be perfectly accept-able until they sat down one day and started to think about what was actually happening. Cara had been spraying urine for some time around the kitchen, both Stripey and Molly were urinating in various places throughout the house and all three females were ill! Molly had been passing blood in her urine and she was in pain when she urinated. Her vet had diagnosed a condition called Feline Idiopathic Cystitis and she was being treated accordingly. Poor Stripey had been diagnosed with chronic renal failure and the vet was suggesting further tests for an overactive thyroid gland. Stripey had also had a couple of episodes of cystitis. Cara was ripping herself to shreds with a flea allergy. Only Egbert remained healthy, in a permanent state of eating and sleeping. Sally and Mark felt they were living in a hospital ward full of sick and depressed cats. How could this situation have arisen? They suddenly realized that they lived in an atmosphere of permanent tension with no two cats in one room at the same time. The quality time they spent with their cats involved washing carpets and walls and popping pills. This wasn't what they had had in mind at all. Their vet sympathized as best he could and suggested they call me.

The facts of the case

It took an eternity to unravel all the intricacies of this case but eventually the important factors emerged.

- All four cats had lived previously as single cats or as strays.
- All four cats were introduced into the household as socially mature adults.
- The three females were equally matched in age and temperament.
- The three females had diagnosed medical conditions.
- Cara was spraying urine indoors.
- Stripey and Molly were urinating inappropriately indoors.
- The cats were provided with one single litter tray in a downstairs cloakroom.

Sometimes I enter multi-cat households that, with the best will in the world, really don't work. Cats are perfectly capable of being sociable together but the species was fundamentally designed as a solitary predator which does not need social co-operation to survive. For any multi-cat group to be successful it is important that the members of the group are compatible personalities and the living space contains sufficient resources to avoid the need for direct competition. Even then, in my opinion, most multi-cat households depend on a degree of mutual tolerance once the members of the group reach social maturity.

Everything about this group reeked of failure. All four cats had lived alone (and liked it) and they all came into the house-hold as mature adults. Their ages and temperaments were similar and therefore none of the three females was prepared to allow one of the others to get the upper paw. Their whole existence revolved around conflict and avoidance, at an age when they should have been toasting themselves in front of the fire and relaxing.

Sadly cat stress does go to the bladder and Feline Idiopathic Cystitis is a condition that is strongly believed to be stress-related. FIC often causes cats to urinate inappropriately in the house so it was hardly surprising that Molly was soiling. Twelve years of age seems to be a bit of a watershed for cats. It is a time when cats often become ill; in my experience if they can live past this dangerous milestone they have a sporting chance of getting to twenty! Stripey had developed chronic renal failure, a condition that doesn't necessarily spell the end but needs to be treated with medication and an appropriate diet. It causes high blood pressure and also usually results in the cat's drinking and urinating a lot as the kidneys fail to con-centrate the urine. Cats with renal problems also tend to get urinary tract infections that can lead to cystitis so it's easy to see how Stripey had also started soiling indoors. Her added complication was a tentative further diagnosis of hyperthyroidism, a condition which results from the growth of a tumour on the thyroid gland and causes drastic changes to the cat's metabolism. It can also affect behaviour and many hyperthyroid cats appear very short-tempered. Even Cara had not escaped with a clean bill of health. She suffered from a flea allergy, which can lead to extreme itching and discomfort as the bite of a single flea can cause a severe dermatological reaction. Fleas are a problem throughout the year and multi-cat households can be breeding grounds for the flea population.

It was important for me to point out to Sally and Mark that any illness suffered by any member of the group could potentially exaggerate problems between them. I was sure that the poor health of the three females was causing even more disharmony. Cara's urine spraying was a typical example of what happens when tension and insecurity reign supreme in a household. Urine spraying is a perfectly normal form of feline communication that provides a strong visual and olfactory signal both for the sprayer and other cats. As I have mentioned, there should be no need for a cat to spray within its home because this should be the one place where it feels safe and secure. Unfortunately, Cara did not feel this strong sense of security and she therefore utilized a behaviour normal in similar states of conflict and arousal. She sprayed urine in all the places where she felt most vulnerable.

Sally and Mark had presented me with an incredibly difficult problem. They were adamant that they didn't want to re-home any of the four cats. They felt that they had taken them on and therefore had a responsibility to make it work.

Whilst that is fine in principle it often proves impossible in practice. I promised my best efforts and proposed the following programme.

The behaviour programme

- Sally and Mark would speak to the referring veterinary surgeon to arrange for Stripey to have the blood test for hyperthyroidism. If this proved positive the vet would treat the condition or perform surgery. The vet would also be consulted regarding treatment for Stripey's kidney failure. (The couple had resisted treatment for the renal problem prior to my visit but I suggested this would probably greatly improve Stripey's quality of life and may even reduce her frequency of urination.)

- Sally and Mark would consult with their vet regarding thorough flea control for the home and all four cats to reduce the likelihood of infestation. (The vet had already recommended this but both Sally and Mark were very keen on alternative therapies and felt that all drugs and treatments were potentially damaging. Whilst this is true in some instances there are always occasions when the benefits greatly outweigh the disadvantages.)

- I suggested a thorough cleaning regime for all the areas that had been soiled with urine. In one particular location there was no alternative but to remove the area of carpet and underlay and replace it. (When this is done it is important to treat the floor underneath to remove any residual odour before re-laying carpet.)

- Sally and Mark were asked to provide three further litter trays in different locations throughout the house. One was positioned in Cara's storeroom, one in the bathroom upstairs and one in a spare bedroom. The original tray remained in the downstairs cloakroom.

- The synthetic facial pheromone spray was used to good effect in the areas where Cara had sprayed urine. I emphasized the importance of removing the urine with just water and surgical spirit. Using any detergents tends to destroy the fragile pheromone and renders the product ineffective.

- Two of the plug-in versions of the synthetic pheromone were fitted to floor-level sockets in the hallway and landing.

- Stripey needed a special diet and therefore mealtimes needed to be closely supervised. The other three cats were placed on diets appropriate for 'senior' cats and special feeding stations were adopted for each cat. Cara loved her storeroom so this was the obvious choice for her. Stripey spent a great deal of her time upstairs in one of the bedrooms so she was fed there, and Molly was fed in another favourite bedroom. Egbert continued to eat in the kitchen, oblivious of the chaos around him.

- Water bowls were provided throughout the house in various locations to ensure there were plenty of opportunities for all the cats to drink.

- Toys and games were included in the family's daily routine to encourage gentle and enjoyable activity. Play is always a good antidote for tension.

- The whole house was assessed to ensure there were plenty of high resting places, private hiding areas (bottoms of wardrobes etc.) and beds to satisfy all four cats and provide plenty of choice.

- Sally and Mark were occasionally making matters worse as they intervened when two cats approached each other. This created further tension in cats and humans that merely fuelled the whole unpleasantness. Sally and Mark were under strict instructions to remain calm and to ignore alter-cations. If there was any risk of injury (none so far) they had permission to create a diversion or distraction, such as a strategically placed cushion between the antagonists or a sudden loud noise.

The outcome

Within a couple of weeks I had the first report from Sally and Mark. As suspected, Stripey was indeed hyperthyroid and the vet had decided to give her tablets to stabilize her condition. All cats were suitably protected against fleas and the house had been treated too. Molly had been put on a senior recipe wet food diet to ensure she was getting enough fluid (water intake is extremely important in cats with Feline Idiopathic Cystitis). All the points of the programme were in place and Sally and Mark were delighted with the progress. Most of all they felt that they had taken control of the situation and they were feel-ing much calmer about everything. There is no question that such a change will always have a positive effect on the cat interaction and there was a definite reduction in the tension. They even reported that Cara's exclusion zone seemed to be getting smaller! The most important result was only one spray of urine and no soiling since the start of the programme.

By the end of the eight-week programme there was another breakthrough. Molly and Stripey had been seen sitting back to back on the same sofa! Sally and Mark knew that to maintain this result they would have to continue to work hard in-definitely. All the elements included in the programme would need to be adopted for life. It takes a lot of effort to create this sort of compromise but Sally and Mark both agreed it was worth it.

Thomas – the case of the stray tomcat

Multi-cat households can be fraught with problems but one scenario I see all too often is the saga of the stray tom. The sort of telephone call I get usually starts like this. 'I recently adopted a lovely cat that had been hanging round for a while and he was so lovely and ever so poorly so I looked after him and got him castrated and my other cats didn't seem to mind him until a few weeks ago when he started to get really nasty towards them and my Lily won't come downstairs now and Bennie's left home and I don't know what to do because the stray, we call him Basher because his ears were all chewed up, is so loving and Lily never sits on my lap . . .' Yes, that really was one sentence! Poor unsuspecting cat lovers take pity on skinny scabby tomcats and they love them and nurture them and, hey presto, look what happens.

Here is a tale that illustrates this phenomenon really well. Carol, John and their daughter Lisa lived in a beautiful and spacious house in a quiet residential area surrounded by trees and fields. They had two cats, Ella and Snowy. Ella was about eleven years old and she had originally lived with the family's previous cat until he died when she was five. Snowy arrived as a companion but the two females never really became bosom buddies. Snowy was always keen to play with her new companion but eventually gave up when she realized that Ella's response was always hostile. Ella also tended to be fairly un-sociable with the family apart from Lisa. She used to sleep with her at night and always went to her when she felt she needed attention. Snowy was slightly different; she was rather nervous and very timid of people as a kitten but she soon learned to trust the family and became very affectionate. Strangers were still a bit of a problem for her but she had her own favourite place, under the bed in Carol and John's bedroom, to retreat to when she felt threatened.

There were no problems in the household until Snowy was about five. During that summer Carol had started to see a very thin and bedraggled cat sitting under the trees at the bottom of the garden. She repeatedly tried to approach him but he always ran away. She became very worried about his state of health so she started to take a bowl of food down to the area where she had seen him to provide nourishment. Sometimes she wouldn't see him for days and then he would return and eat the food with some relish. She kept to a routine and she was delighted to find that, towards the autumn, the little cat would actually be waiting for her under the trees when she took the food down to its normal place. She started to stroke him gingerly when he was eating and her heart melted when she saw how thin and poorly he looked. He had scabs all over his face and a torn ear; his neck and shoulders were broad and his face, despite his weight, was full with thick jowls. As he turned to leave one day after a delicious meal, Carol noticed that he was well endowed in the testicle department and she wondered how he came to be hanging around her garden. She made enquiries with her neighbours and even put a notice in the local

post office but nobody rang to say he belonged to them. She presumed he had been a stray for some time.

Time went on and Thomas (the stray's adoptive name) became very attached to Carol. He was extremely affectionate and friendly and he became terribly excited when he was stroked. He would chew on Carol's hand (ouch!) but she found this endearing and she desperately wanted to integrate him into her home as a third member of the group. The family had a discussion and it was agreed that he should be castrated, given any necessary veterinary treatment and then brought home. The day Thomas returned to his new home he was provided with a cosy bed in the kitchen and a bowl of tasty food. Carol was worried that he would hate being confined and try to escape but, in the morning, she found him curled up in his basket purring quietly. He seemed so passive and gentle that she left the kitchen door open for him to explore the house and hopefully meet his new feline companions. Snowy did a double take when she first saw him and rushed over to sniff him enthusiastically. Thomas appeared to completely ignore her and continued to wander around his new home. Ella was a different story and she hissed and growled at him, and adopted a fierce posture as if she would take his head off. He walked past her with nothing more than a cursory glance. Ella immediately retreated to Lisa's bedroom to dwell on what had just happened.

Carol was delighted; as far as she was concerned she had adopted a gentle pacifist and all would be well. Over the next few weeks Thomas endeared himself to the whole family. He was playful and so intense in his affection (and hand chewing) that he was hard to leave alone. He stuck to Carol like glue and appeared to stare at her with sheer adoration. Isn't this secretly what we all want from a cat? Almost without the family's realizing he became the centre of attention. Snowy tried to play with the newcomer but was either ignored or (probably when no one was looking) punched hard by Thomas and sent flying across the room. Ella was conspicuous by her absence and remained upstairs.

Lisa was the first member of the family to become suspicious. Ella's food and access to outdoors was in the kitchen and she was making no effort to come downstairs to eat or relieve herself. She was leaving the house at dead of night through a small window upstairs and negotiating a treacherous path down a sloping roof onto the garden fence. Lisa was picking her up twice a day to take her downstairs to eat something in the kitchen. She soon realized Ella was not eating well and losing weight. She started to provide food for her in the bedroom but Ella was picking. Instead of sleeping peacefully on Lisa's bed she was seeking refuge underneath it.

Thomas continued to win the hearts of the family but poor Snowy was really getting trounced. She loved to play with ping-pong balls but every time she rushed after one when Thomas was around she would get a beating. Snowy proved to be a feisty little character and she remained relatively undeterred in her movements about the house. It was apparent, however, that she was not entirely happy as she walked slowly with her belly to the ground and made huge detours whenever she saw Thomas. Carol realized that this just wasn't right. Thomas had started sauntering casually upstairs and soon discovered Ella under Lisa's bed. There was a massive fight as Ella, in complete terror, launched an attack to defend the only safe place that remained in the house. Luckily John heard the noise and separated them but he sustained a nasty bite as a result.

The facts of the case

The family decided to take professional advice and I visited them and took loads of notes whilst watching Thomas at work. He was certainly doing his best to be as friendly and cute as possible but I was interested to see the way his face changed when Snowy entered the room. It's amazing how subtle these changes are but, if you know what to look for, they can be very revealing. During the consultation there were many important points noted, the crucial ones being:

- Thomas had a history of straying as an entire tom for an un-known period.
- Ella was not particularly sociable with other cats and had merely tolerated Snowy prior to Thomas's arrival.
- Snowy seemed to be instantly friendly to other cats but was becoming increasingly wary of Thomas. This change in response coincided with an increase in aggressive behaviour towards her.
- Thomas was very affectionate and 'clingy' with the family.
- Ella had retreated upstairs but Thomas was gradually in-vading that part of the house.
- Ella had lost weight.

Thomas had obviously spent a significant period of time living on his wits as a stray. His aggressive, territorial and competitive nature would have been fuelled by his testosterone as he fought for rights of access to hunting grounds and females. Cats surviving in this way can be very opportunistic and many discover that humans provide a regular source of delicious food, as well as living in warm secure environments that beat a cold, damp 'nest' under a bush any day of the week. They can become extremely friendly towards humans but, sadly, this doesn't often extend to their relationships with other cats. Castration obviously removes the hormone that motivates the intense territoriality of the adult male but often this behaviour has become an inherent part of the beast. Thomas was employing an incredibly effective survival strategy. He was passive in his entry into the group whilst he was recuperating and building strength. It also gave him the opportunity to 'recce' the household to see what the competition was like. What he found was a fearful elderly female and a stupid but innocuous five-year-old who was annoying rather than challenging. Neither posed any threat as far as he was concerned. Most of the aggression that Thomas used over the ensuing weeks was covert and passive apart from when he was directly challenged. He relied heavily on stares and psychological intimidation, peppered with the occasional cuff round the head for Snowy just to keep her on her toes. Both Ella and Snowy were showing signs of stress and things were set to become much worse.

I hate being the prophet of doom when I visit clients. Over the years I have come to one very definite but rather frustrating conclusion. Some problems just cannot be fixed. They not only cannot be fixed but they shouldn't be fixed as this would seriously compromise the emotional well-being of the cats in question. This is an important lesson and it's one that every cat owner would do well to embrace. It is, however, very difficult to offer no potential solution whatsoever when I visit clients and there is always a remote chance (such is the nature of the cat) that the impossible will happen. I offered the family my views about the likely outcome to this scenario but suggested a trial period of eight weeks to follow a therapy programme and see what happened. Carol, John and Lisa found it hard to accept that Thomas might have to go. He was such a friendly character and, if they were honest, the most 'user-friendly' of all the three cats. But they knew this wasn't the primary concern and that their loyalty ultimately had to lie with Ella and Snowy.

The behaviour programme

I gave them the following plan with strict instructions to maintain regular contact with me. I was worried the problem would escalate and that a decision would need to be made sooner rather than later.

- Thomas needed to have some rules in his life and although he had been regularly shut into the kitchen at night the family had not always remembered to do so. This now became a strict instruction and he was provided with food, water, a comfy bed and access to the cat flap.
- Ella was to be fed upstairs on a regular basis and Lisa was to devote some time to her in the evenings.
- I asked Carol and Lisa to make some catnip pouches and offer them to Ella and Snowy for short periods every day. Thomas wasn't allowed one (shame!) but it was agreed he was already in heaven so it really wasn't necessary.
- Thomas had to learn that he did not have sole rights of access to the three humans in the family. This is a difficult concept for some cats to grasp but results can be achieved by denying access to the owners on a regular basis. This soon makes them less of an important resource as the cat gets its pleasure elsewhere. The more attention a cat gets, the more it wants. It is also true that the less attention a cat gets, the less it needs.
- Carol was to use cotton gloves to collect facial pheromones from all three cats to spread around the house. I asked her

to stroke the cats firmly around their heads, using the gloves, with particular emphasis on the cheeks, chin and forehead. The impregnated gloves (one to be used per cat) were then scraped against doorways and furniture at cat height to deposit the scent. The process needed to be repeated daily, and was intended to provide a strong and positive scent message to all three cats that there was some sort of communal nature to their relationship. (I wasn't confident in this case but it can work very well.)

- Carol and John were encouraged to play regularly with Snowy and Lisa was asked to do the same with Ella. Play is such a positive pastime and both female cats had become rather fixated on their adversary.

- I asked Lisa to take Ella to the vet to investigate her loss of weight and lack of appetite. This revealed nothing significant so Ella was provided with some gentle flower remedies, in agreement with her vet, that were diluted and added to her food. These remedies were created to treat humans but they have been used in companion animals for many years (see Chapter 2). They are a treatment for negative emotions such as fear, hatred, resentment, and lack of confidence, and there are many specific remedies that are relevant for cats who are experiencing problems. Ella was extremely unhappy and needed a lot of help.

- Carol agreed to make twice weekly reports with a diary of behaviour and interaction.

I tried hard to think of other things that we could do to create an improvement. There are plenty of ways to reduce competition in multi-cat households but all of them were useless in this particular case. There were too many practical complications and the character of the three cats could not be changed dramatically.

The outcome

Carol followed the instructions and the whole family withdrew from Thomas. They were surprised to see that he seemed to accept this quite quickly and started to approach them less and spend more time on his own. Ella enjoyed her play time and the attention from Lisa but she still withdrew to the protective custody of the gap under the bed when Lisa left the room. Snowy loved to play but it was sad to see her looking over her shoulder all the time in anticipation of an attack from Thomas. She didn't seem to feel able to give herself totally to the game. Carol maintained regular reports and they catalogued the deterioration in the household and the gradual realization that it wasn't going to work. The last straw came when poor Ella urinated on Lisa's bed and Thomas sprayed urine on the landing. It was almost as if the cats were coming to the end of their tether and providing their owners with as strong a message of dissent as they could. It was decided that Thomas would benefit greatly from a home of his own and Ella and Snowy could return to normal.

This may seem like a failure in the world of cat behaviour counselling but it's important to know what happened next. Thomas was placed in the safe hands of a local cat charity and it was insisted that a home be found where he could live as an only cat. Ten days later Thomas was travelling to a nearby village with his new owners. They were a middle-aged couple who had just lost their eighteen-year-old cat and they were looking for another companion to share their home. Although Carol wasn't given the identity of the new owners she did receive a call from the rescue cattery a few weeks later. They very kindly informed her that Thomas had settled in very well and was already exploring his new territory. Meanwhile Snowy had quickly returned to normal after a few days of pacing around wondering what Thomas was up to. Where was he hiding? His scent soon faded from the house and she returned to her previously playful ways. Ella took a little longer to realize that Thomas was gone but, after a few weeks, she started to venture downstairs and things began to change for the better.

Choco and Pancake – the case of sibling rivalry

Choco and Pancake were brother and sister and fine examples of brown and cream Burmese respectively. They lived with their owners, Gail and Mike, in a rambling three-storey Victorian house in Essex. When I met them they were seven years old, and until shortly before my visit they had enjoyed a trouble-free existence and a relationship that potentially contradicted everything negative I have ever said about multi-cat households.

On advice from the breeders, Gail and Mike had kept Choco and Pancake indoors ever since they first arrived when they were twelve weeks old. Gail worked a couple of days a week from home so she felt that the company, both feline and human, and the spacious house would provide plenty of entertainment to keep the cats happy. The cats were provided with a single litter tray in the utility room and plenty of toys, but they were mainly active in the evening when Gail played with them. Most of the day they spent fast asleep. They were fed a good quality low-calorie dry food (they were prone to being a little over-weight) and enjoyed an occasional treat of tinned tuna. All things considered, Gail and Mike felt their cats had the perfect life until one fateful day about four weeks before my visit.

Gail returned home one day in the late afternoon and found that Choco and Pancake had been fighting. There were obvious signs of a serious bundle in the utility room; the litter tray was broken and litter was strewn everywhere and the waste bin nearby was on its side. Urine was sprayed all around on the floor and walls. It looked as if something very nasty had happened. Gail found the cats growling and hissing at each other on the landing and she immediately separated them into their own private 'sin bins' so that she could clean up the mess in the utility room and reflect a little on this unprecedented turn of events. Mike returned home and after some discussion they decided to let the two cats out of their rooms to see what happened. Immediately on sight of each other another fierce fight ensued and Mike sustained a nasty injury in his efforts to separate them. The poor couple tried to get the two cats together on three further occasions and fighting began almost immediately. After a great deal of effort they managed to get the cats in the same room but the growling and hissing was distressing to watch. What could possibly have gone wrong when they used to have such a good relationship?

The facts of the case

I listened to Gail's story whilst watching two very disgruntled Burmese eyeballing each other with malevolent intent. I was becoming tense myself because I was convinced that, any moment, I would witness one of their ferocious encounters. Gail told me the following relevant facts during our discussion.

- The cats were kept exclusively indoors.
- The cats had previously had a good relationship.
- They had one litter tray to share.

- Three of the bedroom doors were kept permanently shut because Pancake had a habit of urinating on beds.
- The cats always slept in different locations in the house.
- There were signs of a struggle in the utility room the day the cats started fighting.

When cats live exclusively indoors their whole world exists within the four walls of their home (see Chapter 11). The environment tends to be relatively constant and many cats rarely experience the heady adrenalin rushes that outdoor cats experience on a daily basis. Whatever happened that fateful day will probably never be known but I would guess that one of the Burmese saw another cat outside through the glass door in the utility room. This is pure supposition but Choco or Pancake, on sight of the strange cat, might well have experienced an intense feeling of territorial aggression and adrenalin would have surged through their veins in anticipation of an attack. The other cat may then have entered the utility room to use the facilities in response to a full bladder, and the movement triggered a redirection of the aggression onto the innocent victim. Hence the urine-splashed walls and evidence of a real commotion.

Unfortunately if relationships have not previously been as good as owners have assumed an incident like this can lead to permanent hostility every time one cat catches sight of the other. I didn't doubt Gail's honesty in saying that their previous relationship had been good but she had revealed a couple of facts that may have indicated otherwise. Pancake had a history of urinating on beds; could Choco have been guarding the litter facilities? The cats also slept apart and spent little time together; could this show a relationship of tolerance rather than mutual affection? The added complication was that both cats had experienced some intense emotions over the last four weeks and, in the absence of anything else nearly as rewarding, they were locked into the excitement of battle. They were fluctuating between offensive and defensive gestures when faced with their opponent and they were permanently on red alert. Something had to change to break the cycle.

The behaviour programme

I needed to provide Choco and Pancake with an environment that would give them plenty of entertainment and excitement to provide a healthy distraction from their current obsession with fighting and general aggression. I also needed to address the underlying tension of two cats living in a relatively confined environment who felt they had to compete with each other to survive. Luckily Gail and Mike had already achieved a great deal and, during the times when they were at home, the cats were allowed to mix. They were managing, most of the time, to avoid active fighting. Our aim was to get them back together permanently without the need to separate them when their owners were out.

- I asked Gail and Mike to replace the broken litter tray in the utility room but relocate it to the other side, away from the door, to prevent any future association between its location and a very bad thing's happening.
- I suggested that a further two trays were located in different discreet areas to prevent any possibility of tray guarding if the cats saw lavatories as a scarce resource.
- Two synthetic feline pheromone devices were plugged into sockets on the ground and first floors to give the cats a renewed perception of calm and security.
- Food was distributed throughout the house in six different

feeding stations to give an illusion of plenty despite feeding the same amount.

- Water bowls were provided in every room.
- An entertainment regime began that included a variety of exciting games and toys that Choco and Pancake could explore when Gail and Mike were at work. I also suggested a number of interactive play ideas for evenings and weekends.
- Scratching posts, climbing frames and high resting places were established throughout the house to provide opportunities to escape in every room. Many of these high places utilized existing furniture and shelving – I don't believe that homes have to end up looking like zoo enclosures!
- Any fighting that looked as if it would end in injury was interrupted by Mike or Gail using a pillow but no further intervention on their part was advised. The cats needed to be in control of the outcome of their own altercation.
- After a couple of weeks of this regime Mike and Gail experimented over a weekend by leaving Choco and Pancake alone together for increasing periods whilst they were out of the house.

The outcome

Within six weeks the situation was greatly improved. Pancake was still hissing and growling occasionally, particularly if Choco approached quietly and took her by surprise. Otherwise Gail felt that things were greatly improved and both cats seemed more active and generally happier.

I see a significant number of Burmese siblings falling out as a result of this type of single incident. When I look back at the various cases, an interesting pattern emerges. If the cats fall out around their second birthday (when they would be considered to be socially mature) they are extremely difficult to work with; in most cases the relationship is irretrievably damaged and one of them needs to be re-homed. But if they are over four years old and have a history of co-habiting in 'harmony' before the single incident that starts the fighting, there is every chance that the situation will resolve given time and careful behaviour therapy. Given that the Burmese is one of the most popular pedigrees, even the number of redirected aggression cases I see does not imply that this is anything but an uncommon problem. They are still one of the finest and most intelligent breeds around.

Jemima – the case of 'one too many' cats

Harriet was a real cat lover – so much so that she couldn't resist a sob story. She started her menagerie with a brother-and-sister act called Calico and Charlie. Three years after they arrived as tiny kittens, Harriet heard that a local cat had a litter that the owner didn't want. The rumour was that the fate of these poor kittens was uncertain so she immediately enlisted the help of a group of friends and took all the kittens and distributed them to various homes. She kept two, Jemima and her brother Whiskey. They were very tiny and Harriet lovingly hand-reared them for a number of weeks until they were weaned.

I always feel that stray cats have a sixth sense about the location of homes containing people like Harriet. Three years before I met her, Harriet had noticed a small stray cat hanging round her garden. She duly fed the poor waif and, sure enough, Girlie soon became cat number five.

Harriet admitted that there was always an uneasy truce

between the cats. Calico and Charlie tried to avoid confrontation and spent most of their time hidden away in quiet locations. Jemima, Whiskey and Girlie could be quite antagonistic towards each other and a hierarchy that shifted daily soon became the norm. Prior to a recent house move, the cats had lived in a relatively large family home with access to outdoors via a cat flap. There was no indoor litter tray but all the cats seemed quite happy to eliminate outside.

The problem started shortly after Girlie arrived in the household, when Jemima and Whiskey were about two years old. Jemima started to urinate in the house but Harriet didn't worry too much because the incidence was sporadic and she felt she was on top of the problem. However, this did not remain the case when Harriet and her five cats moved to a new property. She had chosen a rather lovely modern townhouse with three floors but the cats didn't seem so enamoured with her choice. Suddenly all hell was let loose as Jemima persistently urinated just about everywhere apart from the newly acquired indoor litter tray and Whiskey, not to be left out, started to spray urine on every door in the house. By the time I arrived most of the rooms were out of bounds to the cats and the rest of it, consisting of kitchen, hallway, landing and study,

169

was swathed in sheets of polythene and newspaper. It was a real war zone.

The facts of the case

I sat with Harriet as she told me about her situation and every now and then I got a glimpse of one worried little furry face or another. These cats were clearly very unhappy. I paid particular attention to a number of facts that I felt were relevant.

- The referring vet had found that Jemima's urine was very concentrated and contained traces of blood.
- The uneasy relationship between the cats was made worse when they moved to a smaller home with a very different layout.
- The problems started when Jemima and Whiskey matured socially.
- Jemima's soiling and Whiskey's spraying seemed to be worse shortly after a disagreement with Girlie.
- The single indoor litter tray didn't seem to encourage Jemima to use it instead of the floor.

This looked like a classic case of 'one cat too many' in the household. I didn't feel that everything would have been perfect without Girlie since her arrival coincided with the younger siblings' coming of age. It is entirely possible that they would have started getting agitated with just Calico and Charlie around, but my gut feeling was that Girlie was the big issue. Moving to the current house with its distinctive layout was a cat disaster waiting to happen. Each floor was accessed by a narrow staircase, thus creating an environment where an assertive cat could easily control the movements of others around the house. The actual property was much smaller than their previous home and tension would inevitably increase as the members of the group were compressed into a smaller area (think about how you would feel in a crowded lift full of people you either dislike or fear; that's probably our emotional equivalent). Jemima clearly had a chronic cystitis that was causing the soiling. She was uncomfortable and too nervous to go outside; the newly acquired single litter tray was probably being guarded twenty-four hours a day. Even Whiskey decided he'd had enough as his sense of security within the home vanished and he responded in the only way he knew how: by spraying urine. Poor Harriet had to shut doors and protect her lovely wooden floors with polythene and news-paper, merely making matters worse for the confused pussies.

The behaviour programme

I devised the following programme for Harriet but I had to be straight with her right from the start. I was worried about Girlie and her influence over the whole group. Ironically she was the most friendly and affectionate of all the cats, and Harriet reluctantly agreed that she felt Girlie was a huge catalyst. However, she wasn't prepared to give up just yet. So we agreed to try the eight-week programme and monitor it carefully day by day to see what progress and success we could achieve whilst maintaining the current group of five. If we were failing we had to do some serious talking. The basic elements of the programme were as follows:

- The doors to all the rooms were left shut to start with; we had to continue to contain the damage.
- Harriet used synthetic feline pheromones to treat the areas where Whiskey had sprayed and the plug-in version was utilized on every floor of the house.

- The house really needed six litter trays in different locations (using the formula 'one tray per cat plus one') but we just didn't have the space. We compromised and provided four in different locations in the kitchen, the study and an area under the stairs and added a further two adjacent to a couple of the others. It was the best we could do.
- Harriet also provided attractive soil beds in her garden to encourage the more confident individuals to eliminate outdoors.
- Harriet provided a number of high resting places in the kitchen and study to allow the cats to use the available space as efficiently as possible.
- Several secret areas in cupboards were created to allow each cat to have private moments away from the tension.
- Extra beds were provided to try to avoid the conflict that would arise if the cats competed for the same one.
- The cats were put onto a wet diet and given extra sources of water in bowls throughout the house to combat Jemima's urinary tract problems.
- The referring vet prescribed some homoeopathic treatment for both Whiskey and Jemima.

The outcome

I don't think I have ever worked with a more conscientious owner. If anyone was going to make this work it would have been Harriet. She reported to me every other day with exact details of litter tray usage, soiling, spraying and general inter-action between the cats. Almost every day there were diary entries involving some sort of incident with Girlie. All the stuff going on between the cats that Harriet had previously thought irrelevant suddenly seemed sinister and unpleasant. Girlie, without really trying, was sealing her own fate.

Over the next few weeks Harriet made great progress. The extra resources within the home, the wet diet, the homoeopathic remedies and Harriet's determination stopped Whiskey spraying urine and enabled several rooms to be opened to the cats without any incidents. Jemima continued to soil two or three times a week, usually after a run-in with Girlie. Fortunately, she limited her behaviour to two separate locations so Harriet was able to remove most of the polythene and newspaper that had blighted her home for so long. Many owners would have been satisfied with this great improvement and probably made a thousand excuses to keep Girlie – the most people-friendly cat of the entire group. Harriet was made of stronger stuff and she couldn't bear the thought that all her cats were not completely happy. Girlie had to go.

Once again Harriet proved a real winner by endeavouring to find a suitable new home herself. She felt that a rescue centre might view Girlie unfavourably as she had a history of causing conflict in a multi-cat household. Girlie stayed with Harriet and the others for a couple more weeks until, after clever use of the internet, a network of friends and the local paper, prospective new owners came to meet Girlie. It was love at first sight and Girlie went to start her new life as a single cat in a rural idyll in Surrey – thoroughly inspected by Harriet in advance.

At that point I held my breath to see if my theory about Girlie's negative influence on the others was correct. Within days Jemima became a new cat. Harriet reported she seemed more relaxed and more affectionate than she had ever been before. The most incredible result was that from that day forward Jemima never soiled in the house again.

More and more cats are being kept in situations that clearly represent social overcrowding both inside the house and within the territory. The confusing factor is that the concept of overcrowding is an individual thing and some cats are more tolerant than others. Despite my reservations in principle, it is completely unrealistic to presume that multi-cat households will ever become a thing of the past. There are just too many cats to go around and the alternative of thousands more homeless cats is too horrific to contemplate. The best compromise therefore is to try to understand our pet cats better and minimize the potential distress of group living.

CHAPTER 8

Anxiety and Fear

ALTHOUGH I HAVE CHOSEN TO DEAL WITH FEAR AND ANXIETY in the same chapter they are emotions that manifest themselves in different ways. A cat can be anxious with no outward signs, particularly if he has an introvert personality; owners often seem amazed if I suggest that problem behaviour is resulting from their cat's underlying anxiety. Identifying an anxious cat involves watching for subtle changes in routines and patterns of behaviour or other indications (see below) that all is not right. Anxiety can be acute and related to a specific event such as a visit to the cattery, the arrival of a new cat on the block or redecoration in the home. It can also be chronic and caused by unresolved stress triggers in the environment, poor early socialization or an underlying innate nervousness that resists all attempts to habituate to everyday challenges. Signs of anxiety include

- tense body
- lip licking
- dilated pupils
- hypersensitivity to noise/movement/touch
- urine retention
- urine spraying
- inappropriate urination
- over-grooming
- change in normal routine/patterns of behaviour

Fear tends to be rather more obvious and occurs in response to something that the cat perceives as dangerous. Fear of something genuinely dangerous is entirely appropriate but occasionally cats become fearful in situations that don't merit such a response, either as a result of a single traumatic incident or from an underlying inability to regulate the amount of fear experienced in the presence of challenging situations. Signs of fear include

- dilated pupils
- rapid respiration
- rapid heart rate
- tense or rigid body
- mouth breathing
- low crouched body posture
- piloerection (raised fur on the back and tail)
- sweaty paws
- trembling
- aggression
- escape
- hiding/freezing/avoiding
- involuntary elimination
- anal gland expression

❧ ❧ ❧

I am often asked to see shy or nervous cats with a view to improving their quality of life and making them less fearful. I wish it was that simple. Many cats have an innate timidity and expectations have to be realistic. These cats may well adjust to the presence of familiar people and regular routine but anything else would probably still cause them to run and hide.

If you are living with a shy or timid cat you will already know how frustrating it can be. No amount of love or comfort seems to make a difference and your cat continues to disappear under the bed at the drop of a saucepan lid. Many potential owners visit rescue catteries and intentionally choose the frightened cat cowering in the corner. We cannot resist a cat who looks in need of a caring home, but a note of caution here. The cat may be responding fearfully to the strange environment and blossom as soon as he gets home. However, he could just be nervous by nature, so a few enquiries about his background may yield important information.

If you chose a kitten based on a decision of the heart rather than the head it is possible you came home with the one who was hiding behind the chair because you felt sorry for him. Now, several years down the line, that same cat is still hiding behind chairs and you are wondering what went wrong.

These cats will never be laid-back companions but there are certain steps that you can take to ensure that their lives are as good as possible. Many of the actions we take merely reinforce their fear and anxiety rather than make things easier for them.

Tips for coping with an anxious cat

- Do not stare at the cat since direct eye contact is challenging in cat language.
- Try wherever possible to stick to a routine and remember that any environmental changes are potentially challenging and frightening.
- Try to encourage your cat to indulge in the natural behaviour of hunting by using fishing-rod toys or other objects that simulate prey. Even the most nervous cat can lose himself in play occasionally.
- If your cat has a good appetite or enjoys a particular food treat try to encourage him to eat it from your hand to increase the human/cat bond. This should be done in moderation to ensure you do not end up with a fat cat!
- Allow your cat to seek out hiding places where he feels safe and try not to disturb him whilst he's there.
- Do not actively try to get him to face new things by enforcing your will. Exposure to new experiences should appear to be his choice rather than yours.
- If your cat has shown fear in response to a particular object or sound then try to expose him to the same thing to a lesser degree – for example, a quieter version of the same sound. You can then increase the volume until he gets used to it and no longer responds fearfully. 'Flooding' the phobia with constant full-on exposure just won't work.

- Understand that interaction with this cat may not be in the usual tactile way and could involve instead gentle kind words or even games.
- Getting another cat who is less fearful to try to promote confidence by example probably won't work. Many nervous cats are frightened of other cats as well as everything else and this could be their worst nightmare.

Flower essences and herbal remedies

If your veterinary surgeon is in agreement it may be helpful to use certain flower or herbal remedies as a safe and gentle aid in the treatment of anxiety. Skullcap and valerian is available as a veterinary preparation in tablet form and is indicated for travel, urine spraying and general anxiety. Certain flower essences are indicated for fear and anxiety and these can be very effective.

Aspen This remedy is useful for treating cats who were probably born nervous. It's also very good for the jumpy ones who panic when anything new or unexpected happens. I've also used aspen for cats who are frightened of going outside or who wet themselves when bullied by another cat.

Cerato This remedy addresses lack of confidence as part of a cat's general character. It is particularly useful for cats who become dependent on their owners as a result of their lack of self-assurance.

Larch This remedy is very useful for cats who have lost confidence or who are easily intimidated by other cats.

Mimulus This remedy is good to treat cats who have a very specific fear such as other cats or car journeys. It's also very good for the generally timid cat.

Rock Rose This remedy forms part of the combination treatment called Rescue Remedy. Let's hope you never have to use this one because it is recommended for cats who are so terrified that they could harm themselves in trying to escape a perceived danger.

Walnut This remedy is not specifically for anxiety but it is great for treating cats who find it hard to adjust to new circumstances or environments. Timid cats love routine and any changes to it can affect them profoundly.

When you have found the remedies that you feel best suit your cat's emotion it is important to dilute them before administration. Two drops of each chosen remedy should be added to a 30ml container of spring water and four drops of the resulting solution given four times a day, added to food or water or placed directly on the tongue in the more compliant patient. The solution will remain fresh for up to three weeks.

Here are three cases that illustrate different presentations of fear and anxiety that may strike a chord with you. In each situation you will notice a common recommendation that is useful when dealing with all anxious cats. Ignore them!

Gus – the case of the scaredy cat

Gus was about five years old when I visited his home. I understood from his owners, Jennifer and Peter, that he was a handsome lilac Oriental but I had to take their word for it. During the consultation I only ever saw fleeting glances of him as he navigated his way through the house at great speed, moving from one hiding place to another. Poor Gus was nervous. He spent his life in a constant state of red alert, hiding from imaginary demons. I cannot think of anything more exhausting.

Jennifer had purchased Gus from a breeder when he was about two years old. The fact that he was available for sale was motivated by the breeder's apparent desire to 'change the gene pool'. Maybe this statement alone should have raised alarm bells but Jennifer was not deterred, even when she witnessed the breeder's vain attempts to capture and hold Gus during her visit. When he was eventually pinned down and manhandled into a cat basket he travelled to his new home with Jennifer and Peter, but from that day forward things didn't really work out according to plan. Jennifer had always understood he was nervous but considered this to spring from a general dislike of his previous home. She believed that with a mixture of love and patience Gus would become a happy and relaxed (and extremely grateful) pet. However, despite Jennifer's best efforts Gus remained an elusive night-time prowler who disappeared during the day into quiet corners away from the activity in the home. Many of the couple's friends were not even sure that Gus existed. There were times when Jennifer felt the same since she had not touched Gus once from the day he arrived. Something had to change.

There were all sorts of reasons why Jennifer had waited three years to call me. The arrival of her young son, Joshua, had been relatively time-consuming and she had only recently discovered that there actually were people like me who dealt with such things as nervous cats. So there I was, five years down the line, working with a patient I couldn't even see.

Gus was kept as an indoor cat and Jennifer and her husband had created an elaborate routine over the past three years to avoid causing him any unnecessary stress. He thrived on order and predictable activity and it was possible to set your watch by his patterns of behaviour. He would enter the kitchen at a certain time to eat some food but only if Jennifer remained perfectly still and didn't look at him. He would play at night if his toy was placed in a particular location in the evening. He had progressed sufficiently to enter the living room when his owners were sitting quietly watching television but he only remained if they kept perfectly still. It was fascinating watching Jennifer walking in her own home. She had, quite subconsciously, developed a way of moving around that can only be described as 'pussy-footing'. I started to wonder whether the origin of this phrase actually related to an owner's behaviour in response to a nervous cat rather than, as is generally believed, the cat's own distinctive walk.

Jennifer was pleased that Gus didn't seem overly traumatized when Joshua arrived. She was often astounded to find him alone in the room with the baby just staring at him, despite the screaming. Poor Jennifer felt that she had achieved something over the past few years, but now Gus's relaxation seemed to have reached a level beyond which he couldn't go.

The facts of the case

As always there were clues to Gus's behaviour that were extremely telling:

- The breeder had kept Gus until he was socially mature but then considered him unsuitable for breeding.
- Jennifer used to talk and look at Gus constantly whenever she saw him or knew where he was to encourage him to interact with her.
- Gus seemed more relaxed when his owners were still or ignoring him.
- He loved routine.
- He would play boisterously at night when the house was dark and his owners were asleep.

Gus was the sort of cat who was born anxious. He produced too intense a response to everything vaguely challenging and would have missed out on a great many early experiences because he was just too scared to face anything. As a result he was suspicious of people and suspicious of change. Many cats like Gus cope with the stresses of life by living within strict routines that are predictable and therefore considered safe. Any deviation from the norm is potentially a problem and the cat withdraws to a quiet place until familiarity and routine are restored. Gus was using up his reservoir of energy at night when the humans were not around and he had the environment to himself.

An owner's instinct is to persistently touch, stroke and pick up a nervous or antisocial cat to teach him that no harm will arise from contact with humans. Whilst this works with a certain type of cat it certainly wasn't working for Gus. Any attempts to touch him were met with a rapid withdrawal (he always put himself in a location where escape was easy when his owners were around) and the constant talking and focusing were just not getting the message through that people are great. Unfortunately Gus interpreted Jennifer's actions as challenging and distressing and they merely reinforced his dislike. On a positive note, after three years, Gus's owners had become a familiar part of his environment so we did have something to work with.

The behaviour programme

Cats like Gus always respond better to interaction that is more sympathetic to the feline social signalling. Gus just didn't understand that to a human direct eye contact is a good thing; a good eyeballing from next-door's cat is hostile and threatening. Jennifer and Peter had to learn to behave differently around Gus if they wanted the relationship to change. We were going to focus on play and positive interaction that appeared to Gus to be inviting and non-threatening. This was the plan.

- Jennifer and Peter had to ignore Gus. No more eye contact, verbal communication or direct approaches. If Gus entered the room they were to focus on something else and pretend he wasn't there.
- I wanted to use food as a treat that would really motivate Gus but there was always a large amount of food available to him throughout the day. I needed him to be a little hungry rather than wandering round with a permanently satiated appetite. I asked Jennifer to cut down his tinned food by half and weigh out the dry biscuits carefully to ensure he was getting a rationed supply.
- The dry food was placed in bowls throughout the house with a little food foraging from egg boxes and cardboard tubes thrown in to build confidence.
- Gus enjoyed a treat occasionally of cooked chicken so, every night, a small saucer of chicken was placed on the floor near where Jennifer and Peter were sitting in the living room. If Gus wanted the chicken he had to risk walking close to his owners. Jennifer and Peter were asked not to praise Gus if he ate the food because praise from them wasn't much of a reward to this particular cat.
- I recommended the use of certain Bach flower remedies (after discussion with their vet) including mimulus and aspen to help to address the fear that Gus was experiencing.
- Peter was to bring home cardboard boxes and novel items for Gus to explore to try to introduce more challenge to his day-to-day life. Each box should contain a treat of biscuits, toys or catnip (a particular favourite with Gus).

- Over the eight-week period of the programme, Jennifer and Peter were asked to place the bowl of chicken nearer and nearer to their seat until it was actually placed on the sofa cushion beside them. Once this was readily accepted the next stage involved placing the chicken on a flat hand resting on the cushion. This really would be a challenge for Gus.
- Gradually the numerous hiding places within the home were blocked up until Gus was left with only a handful of secret locations. These are an essential element of all cats' lives; it's important to have time out. We just didn't want Gus to spend his whole day hiding.
- Jennifer and Peter took it in turn to play with Gus with a long fishing rod with a feather on the end of a piece of string. Gus couldn't possibly resist this for too long.
- I asked Jennifer and Peter to stop 'pussy-footing' around and walk about their home normally, making as much noise as they wanted. Gus had to get used to a normal household and if his owners were more relaxed it would provide a good positive signal.

My final instruction to Jennifer and Peter was 'patience'. They had a lifetime of Gus's fears and phobias to overcome and any significant improvement would take time. I awaited their first report with great anticipation.

The outcome

After one week Jennifer called to say that all the parts of the programme had been started. Gus had found his biscuits already and she felt (observing him out of the corner of her eye) that he appeared more relaxed. The saucer of chicken had proved to be a bit of a problem for him and he resisted for two days. By the third he decided his love for chicken outweighed any potential danger and he ate it all. Jennifer had noted that from that day forward it took less and less time for Gus to make the decision to approach the saucer; she was delighted.

By the second week the feather on the fishing rod was an enormous success and Gus's initial self-conscious attempts to play were replaced by athletic leaps and twists in the air. Every week seemed to show some improvement and Gus was eating off Jennifer's hand by the eighth week. We were keen to introduce handling and we used a rod with a soft pad at the end to get him comfortable with the sensation of touch. Jennifer held the rod and gently stroked the soft pad across his back when he was eating or relaxed. Eventually she placed her hand further down the rod so that her hand wasn't far away as the pad touched Gus's coat. Within a couple of months Gus was sitting on the back of the sofa next to Peter, lying relaxed with Joshua, purring and generally behaving more like a family pet. Jennifer was still struggling with the handling issue but continued patience often reaps rewards in the end. She was certainly happy with the progress that Gus had made and she soon got used to the new technique of interaction with him.

Squeak – the case of the husband-hater

Squeak (so called because of her rather ineffectual miaow) was a little six-year-old black and white cat who had lived all her life with Sarah. She had always been a rather nervous little soul, even when she was a tiny kitten, but with the help of a gentle owner and a predictable routine she coped very well. It took her about three years to be brave enough to go outdoors, preferring the relative safety of sleep indoors to while away the hours. Once she discovered the delights of the garden she would occasionally venture as far as the bottom fence! It might not sound much, but it was a big thing for Squeak.

When Squeak was four years old, Sarah fell in love with Kevin and eighteen months later they were married and Kevin moved into Sarah's house. Sadly that was when the problem started. Squeak was always wary of strangers and her strategy was to hide and keep quiet until they left. When Kevin used to stay for weekends before he moved in she would merely disappear until Monday. Kevin was convinced for some time that Squeak was a figment of Sarah's imagination because he didn't even see her for nearly a year. Squeak could just about cope with the visits but when Kevin moved in (complete with his 'stuff') she was not a happy cat. Kevin tried extremely hard to win her over; he was smart enough to know that the way to a woman's heart is to pal up with her cat. Unfortunately he tried a little too hard. He thought that it would be a good idea to pay Squeak a lot of attention so he stared at her, talked to her and rushed towards her to greet her when he came home from work. Squeak was appalled. Kevin then decided that he would play with Squeak with her favourite feather-on-a-stick toy. Sadly he got rather enthusiastic and waved it around with such gusto that Squeak was scared witless and disappeared for two days. The *pièce de résistance* came when Kevin looked after Squeak when Sarah went away for a couple of days. What actually happened then is still a mystery but Sarah returned to find piles of cat faeces and pools of urine everywhere, an agitated Kevin and an even more agitated Squeak. That was the day Sarah and Kevin had their first row and it was decided then and there that they needed help. Kevin and Squeak not getting on was not an option.

The facts of the case

I found both Sarah and Kevin instantly likeable and I felt incredibly sorry that all Kevin's well-meaning efforts had been in vain. There was no question that Squeak was frightened of him and her strategy was one of complete avoidance at all times (very hard when Kevin constantly tried to make friends with her). During the consultation they were both open and honest but even I couldn't get to the bottom of the events during the few days Sarah was away. I decided not to pursue it and merely noted that 'something really scary happened'. The important facts were these:

- Squeak had always been nervous and timid, therefore it was safe to presume this was part of her fundamental character.
- She thrived on routine and only really felt comfortable with Sarah, a quietly spoken and feminine lady.
- Kevin was a very tall and muscular man with a deep voice, very different from Squeak's idea of a non-threatening human (Sarah).
- Kevin had tried to win her affection by focusing on her and approaching her frequently.
- Kevin didn't know the meaning of restraint when it came to playing with cats.
- Kevin was frustrated and upset that Squeak rejected him.
- Squeak was spending a lot of time under a bush in the garden since the few days alone with Kevin.
- She would often stay out all night if Kevin came home first because she was too scared to enter the house when he was there.

Just like Gus the lilac Oriental, Squeak had probably not benefited from early socialization so humans appeared threatening unless familiar. Squeak had learned to trust Sarah, whose behaviour and body language were very gentle and feminine. In contrast Kevin could easily have been another species and he obviously appeared very frightening. His focusing on Squeak would have represented a threat and the approaches would look dangerous to her, requiring an immediate retreat. Whatever happened during those few days alone together, it is clear that Squeak was very upset if she defecated and urinated on the floor in various places rather than in her litter tray. The incident would have confirmed in her mind that Kevin was not to be trusted and was to be avoided at all costs, even if it meant staying outside and away from Sarah.

The behaviour programme

The programme focused almost exclusively on Kevin and he was willing to do anything (in theory) to restore relative harmony in the home. I suggested the following.

- Kevin was asked to ignore Squeak, not to talk to her, look at her or approach her.
- A plug-in synthetic pheromone device was recommended to signal calmness and security.
- Kevin and Sarah were to arrange their routine in the evening to ensure that Sarah always came home first. This way Squeak could be called in from the garden and enter the house without worrying about bumping into Kevin.
- Sarah was going to reduce the amount of food put down to ensure that Squeak remained a little hungry (just like Gus).
- Kevin should start to provide her with her biscuits twice a day in the kitchen.
- Every evening Kevin should place a small bowl of treat food (Squeak's favourites were cheese, ham and chicken)

near to Squeak. He would not look at her and would always approach the location without walking directly towards Squeak. He would then quietly leave the room.

- If Squeak remained relaxed then Kevin could repeat the process with other small amounts of treat food in the same evening.

- Kevin should also try offering Squeak's favourite treat by hand but not persevere if she appeared frightened.

- If Squeak retreated to another room then both Sarah and Kevin could also spend time in that room without paying any attention to her.

- If Squeak showed fear when confronted with Kevin then Sarah was instructed not to comfort her. This could reinforce the fear.

- Squeak's favourite bolt-holes should remain accessible at all times so that she didn't panic if she found a door closed. She should not be disturbed whilst hiding in any secret places.

- Kevin should try to act normally whilst walking around the house; he was currently approaching every room with trepidation in case Squeak was in there. I suggested he behave in a quiet and relaxed fashion but took away the air of caution.

The outcome

Within a matter of days Squeak was taking treats directly from Kevin's hand. She was walking nonchalantly past him in the hallway and sitting watching him during the evening as he took great pains not to watch her. Not for the first time I was accused of switching cats during my visit because of the dramatic instant result. I recommended caution with the presumption that things had resolved so quickly and, sure enough, there were several setbacks over the next few weeks. Kevin kept following the programme and, in his words, felt at times that he would be the one in need of therapy before the end of the two months. However, he survived and reported that it had been a great lesson in the art of patience and dealing with frustration. I wouldn't say that Squeak eventually became Kevin's bosom buddy but they co-habited nicely and Squeak soon learned that, despite the obvious differences between him and Sarah, Kevin wasn't that dangerous after all.

China – the case of the needy cat

China was five years old when I met her and she lived with a young couple called Sue and Joe. China was a slightly rotund black short-haired cat and her owners doted on her. She had been obtained from the local cat rescue centre the previous year. Sue and Joe had chosen her because of her incredibly sad story. She had lived with three previous owners; each time she had been returned to the cattery before being re-homed again. The original owners gave her up because she had lived with other cats and they felt she was the victim of bullying. Both the subsequent owners took her back because she soiled in the house. Sue and Joe were apparently undeterred – after all, it was a sad story – and they signed the paperwork and brought her home vowing that this would be the one within which she would spend the rest of her days.

China settled in really well and soon became attached to Sue, following her everywhere. Sue was aware that she had been bullied in her original home so she decided that China would only go outside when her owners were around to ensure the neighbourhood cats kept away. After work Sue would come home and spend a short period of time in the garden whilst China had a patrol round the perimeter fence, content in the knowledge that she was protected by Sue. Occasionally Sue would come indoors briefly and leave China outside. Any loud noise or sudden movement would result in a speeding cat's bursting through the back door and hiding behind her owner or leaping onto her lap. Sue would comfort her and reassure her that she was safe.

The relationship rumbled on with Sue and Joe speaking kind words and stroking and fussing over China. They felt that they needed to compensate for a great deal of hardship in her previous life and they intended to spoil her. One day, over a trivial matter, Sue and Joe had an argument. Sue remembers becoming quite agitated and shouting in the bedroom. Both Sue and Joe were reduced to a stunned silence when China (who had been pacing and miaowing on the landing) suddenly jumped onto the bed and peed on the duvet. She'd never done that before but the memory of her past history suddenly came back to haunt them.

She was subsequently diagnosed with a chronic cystitis problem that was aggravated by stress and her vet suggested they contact me.

The facts of the case

Feline Idiopathic Cystitis is caused by a number of factors but there is a significant stress component to all cases (see Chapter 5 for further details). Episodes tend to be self-limiting and often the only sign that owners will see when their cats are ill is a sudden soiling problem in the house. Cats will strain or urinate frequently and pass blood. A diagnosis is made once other possibilities have been eliminated, including bladder stones, crystals and infection. The important point here was the stress that China was experiencing and what was causing it.

- China had an unfortunate history of being re-homed several times.
- It was understood that China had been bullied in her first home and it was apparent she was scared of other cats.
- There were quite a few cats in the neighbourhood and many used China's garden as a thoroughfare.
- China had become very reliant on Sue and went to her for comfort when she was frightened.
- Sue comforted and reassured her repeatedly.
- Sue spent a great deal of time stroking her and cuddling her.
- Sue and Joe were planning a family.
- Sue and Joe were not giving up on China and certainly not re-homing her.
- China always preferred to come indoors to use her tray rather than toilet outside.
- The litter tray was located near the glass-panelled back door.

At last China had a home where the owners were prepared to deal with the problem rather than give up at the first sign of trouble. Keeping a cat with a behavioural problem isn't necessarily always a good thing since many object to a specific element of their environment and their stresses resolve when they are re-homed. However, in this case the home was potentially perfect for China; we just needed to make a few adjustments.

Sue had looked after and loved China in the best way any-one could. She comforted her and protected her but, as a result, she encouraged China to become totally dependent. China would immediately withdraw from any challenging noise or sight and seek out Sue. Her owner would then reassure her and, by doing so, confirm that the suspicious noise or sight was indeed dangerous. China wasn't learning any self-reliance at all and when Sue started shouting and appearing agitated it was too much for China to bear when she picked up on the vibes. Urinary tract problems have stress triggers and the build-up of tension between Sue and Joe and the subsequent argument may have tipped the scales for China.

There were also issues with other cats outside and a general sense that everything was frightening. It's hardly surprising that China felt overwhelmed from time to time. I was also concerned that the arrival of a new baby in the household would herald a terrible bout of cystitis, just at the time when Sue would have other things on her mind. We had to plan ahead.

The behaviour programme

- I asked Sue to start interacting differently with China. If China showed fear inappropriately then Sue should ignore her. If China appeared relaxed in the presence of novel sounds and objects then she should be praised and stroked.
- China should be encouraged to go outside without Sue, so the back door was left open when Sue was at home.
- The vet had already recommended a wet diet to increase China's fluid intake to help reduce the risk of cystitis. A small amount of the dry food that China was previously fed was placed inside cardboard egg boxes to provide her with a challenging game with food.
- There were no secret hiding places for China in the house and it was important that they should be provided. Sanctuary could represent a suitable alternative to seeking comfort from Sue when China felt vulnerable. Three private spaces were created under a spare bed, in a wardrobe and at the bottom of the airing cupboard. Sue

and Joe were instructed not to disturb China while she was there and to behave normally to signal the obvious lack of danger.

- The litter tray had previously been situated in a vulnerable location in full view of invading cats so it was moved to a discreet corner of the kitchen well away from the door. Another tray was also placed in the upstairs bathroom.

- Sue and Joe were asked to interact more with China by using play. I suggested they acquire some fishing-rod toys and try to encourage China to follow the moving objects.

- China loved to be groomed so this continued since both owner and cat found it very relaxing.

- Sue had several friends with young children and I suggested that they visited regularly for short periods. If the children screamed or cried then Sue and her friend should not react or look at China. If China decided to hide in her new places that was fine. Sue would feed China immediately after the visit from the babies with a small amount of her favourite food (tuna).

- I asked Sue to ignore some of the many approaches China made to her every day. This is a difficult thing to ask because it feels so much like rejection but it was important to encourage a degree of self-reliance and an understanding that there were things to do outside the relationship.

The outcome

During the next few weeks China had a number of crises including a barking dog, a cat on the shed next door, a road drill and three visits from noisy children. Sue remained steadfast and refused to comfort her for such non-threatening occurrences. China appeared to love the play sessions and approached them with great gusto although she did occasionally get so carried away that she frightened herself and sought refuge upstairs. She loved her new litter tray in the bathroom and used it regularly every day, virtually ignoring the original one in the kitchen despite its new location. Sue reported that China seemed more relaxed which was a great relief since she had presumed the lack of attention at those difficult times would upset her. China soon learned that if Sue wasn't frightened then there was no real need for her to be either. On those occasions when she felt it was necessary, she relished the delights of the private dark places where she could hide away from the danger of invading cats and all things scary. Apart from one indiscretion in the bath China didn't soil indoors again and showed no further evidence of cystitis. Children still represented a problem for her but, with the security of the dark places, I felt that this was not an insurmountable problem. She started to spend more time on her own exploring the garden and Sue felt she had certainly acquired the self-reliance we were striving for.

CHAPTER 9

Medical/Neurological Problems

THERE ARE ALWAYS GOING TO BE THOSE CASES THAT WILL NOT respond to a DIY approach. A significant amount of undesirable behaviour has its origin in physical disease and tackling it on a purely emotional level is pointless and potentially delaying important veterinary intervention. This chapter will explore a number of problems that are commonly seen and hopefully give you a better insight into their cause and possible treatment.

Over-grooming

There are a number of different techniques of over-grooming;

all represent part of the normal sequence of coat maintenance. Some cats will lick repeatedly in one particular area at a time and this will cause alopecia (hair loss) and occasionally trauma to the underlying skin. Others will pull at clumps of fur (fur-plucking) and remove great tufts each time. Whatever method is used, the behaviour has one thing in common. It is performed to an excessive degree and it causes loss of fur and a great deal of concern to the owner. A diagnosis of psychogenic alopecia or dermatitis is often made but in reality over-grooming is more likely to have a physical cause. It is possible for over-grooming to be purely stress-related (Parsnip's case below is an illustration) but it's essential to have a full veterinary investigation first before disappearing down the psychological road.

The most common causes of over-grooming are
- fleas and other parasites
- fungal infection (dermatophytosis)
- allergy
- seborrhoeic dermatitis (malassezia)
- localized pain

If your cat has recently started to over-groom it is worth asking yourself a number of questions. Has there been a major change in the routine of the household recently? Have you had any major building work done or redesigned the garden? Have you introduced a new cat? Has a member of the cat group died recently? Are there any new cats on the block? How often do you use flea control?

It is no longer sufficient to use flea control just in the summer months; fleas are there potentially all year round. An adult flea visiting your cat will lay eggs in your carpet. Over a period of a couple of weeks these eggs will develop through larval and pupal stages and emerge, potentially, as an awful lot of fleas in your home. The most effective treatment involves environmental sprays as well as 'spot-on' applications that are deposited directly onto your cat's skin at the back of the neck. If you are not treating all the cats in the household (and any dogs) then it is possible that your pet's problem may be flea-related. A visit to your vet and a possible referral to a dermatologist will diagnose or rule out any medical cause for the condition, including the presence of any potential pain. If nothing can be found then it is worth doing a bit of detective work to see what could possibly be causing your cat to be so stressed.

Parsnip – the case of the stressed over-groomer

Parsnip was a sweet little long-haired tabby and white cat. When I met her she was ten years old and she had lived all her life with her owner, Hilary, and an assortment of other cats and dogs over the years.

When they first moved to their beautiful rural home four years previously, the house had been a relative menagerie of pets. There were two bruiser cats, a brother and sister act called Fred and Ginger, and a frail old dog called Peggy. Parsnip and Peggy were inseparable and they would often be found curled up asleep together. Fred and Ginger actually belonged to Hilary's grown-up daughter but Hilary had agreed to be their custodian during her time at university. Parsnip didn't really have much to do with Fred and Ginger but their relationship wasn't that bad. The siblings spent a great deal of time outside and Parsnip was more home-loving, preferring to sun herself within easy reach of the house.

Life was pretty idyllic until about a year before my visit

when Parsnip lost her constant companion, Peggy, to cancer. She seemed really depressed for a while and it was difficult enough for Hilary to deal with her own grief without worrying about her little tabby and white cat. A further major upheaval took place when Hilary's daughter established herself in a new place and requested the return of Fred and Ginger. Removing cats that are merely tolerated is not the end of the world in theory but it did represent a massive change. Fred and Ginger had been great custodians of the territory and it didn't take long for the local thuggery to realize that Parsnip was fair game. The garden and surrounding woodland were suddenly full of sinister felines with their eyes firmly fixed on the house. The most dangerous and confident of all appeared to be the neighbour's white Persian, rather paradoxically (and absurdly) called Mummy's Little Tulip.

Hilary had found her boyfriend, Duncan, a great source of

strength and comfort when Peggy died and their relationship blossomed. It was decided that Duncan would move in with Hilary a few months after Fred and Ginger had left. Within weeks, however, Hilary noticed that Parsnip wasn't right at all. She had started to vomit regularly and her coat was missing in huge patches on her sides and stomach. A visit to the veterinary surgeons had found no obvious medical reason for this problem and they suggested she was over-grooming due to stress. Parsnip was prescribed Valium but Hilary didn't persevere with the medication when she saw her beloved cat go all wobbly and fall off the furniture after the first few tablets! The vet suggested she speak to me and that's how I found myself in her kitchen, listening to this story, with a little tabby and white cat sitting next to me.

The facts of the case

Once again there were a number of very important comments made by Hilary during our discussion.

- Parsnip had become more dependent and vocal since the death of Peggy.
- Mummy's Little Tulip had entered the house several times through the kitchen window since Fred and Ginger had moved out.
- Parsnip had always slept on Hilary's bed with her but Duncan had insisted that she was shut out at night once he moved in because she made him sneeze.

I agreed wholeheartedly with the referring vet that Parsnip was indeed stressed. Cats can adopt a variety of coping strategies when times are difficult and they often have a basis in normal behaviour, such as grooming, urine spraying or eating. When Peggy died Parsnip turned to her owner for support and reassurance; her increased vocalizing was another indication that things were causing her some anxiety. The removal of Fred and Ginger meant that her territory lost its defenders and cats are quick to capitalize on this. Mummy's Little Tulip (and his cohorts) would have patrolled the range regularly and the lack of the relevant scent would soon have spread the word. The area was up for grabs and the only remaining hurdle was a small and ineffectual tabby and white cat. It was bad enough when Parsnip knew the Persian was in the garden and staring through the window, but the ultimate nightmare had come true when he started to venture into the kitchen and eat her food. Poor Parsnip could no longer turn to Hilary at night for comfort (the last straw) and she desperately needed an outlet for this anxiety. Her over-grooming had a 'stress-busting' effect on her that she would have found comforting and pretty compulsive. The down side was persistent vomiting due to the accumulation of numerous furballs and large patches of alopecia as she simply licked and chewed all her fur away.

Valium is often prescribed by veterinary surgeons to combat anxiety in cats but it is not necessarily the most effective treatment in every case. Vets are limited in their choice of psychotropic drugs because so few are actually licensed for use in cats. Unfortunately the response of cats to diazepam is very variable and they will often become uncoordinated. The vet in question was extremely receptive to other ideas regarding medication and was willing to try more specific drugs, with the owner's consent, to combat the problem in conjunction with behaviour therapy.

The behaviour programme

We desperately needed to restore the relationship between Parsnip and Hilary without encouraging dependency. Parsnip also had a very strong sense of insecurity and inability to defend her territory and it was essential to try to create a safer atmosphere indoors. It is not always necessary to recommend the use of drugs with over-grooming problems but often the cat's brain chemistry influences the intensity of the behaviour and things need to get back into balance internally.

- I spoke to the referring veterinary surgeon about the use of a certain tricyclic antidepressant. I provided her with a number of veterinary papers regarding its use for this sort of problem and she was keen to give it a try. Pet behaviour counsellors are not qualified to prescribe drugs (unless they are veterinary surgeons themselves); the decision is always made by the referring vet. It was agreed that a twelve-week course of medication would start, with a gradual reduction of dosage at the end of the period. I warned Hilary of the potential side effects in the first few days to avoid any risk of the medication's being withdrawn unnecessarily.

- Parsnip's diet was changed gradually from a mixture of wet and dry to a complete veterinary formulated dry food that had an additional element to combat the formation of fur-balls. It may appear that I do this just for the hell of it some-times but I wanted to give Parsnip some entertainment and distraction indoors and 'food foraging' is always a useful tool. You just can't hide tinned salmon in jelly inside toilet roll tubes! Very messy. Hilary was advised to get Parsnip used to eating in different locations and then start to make the acquisition of the food more challenging.

- I emphasized the need for increased fluid intake so

additional water bowls were placed throughout the house.

- I recommended regular flea control for Parsnip to prevent any visitors on her coat making the over-grooming worse.

- Hilary was to shut all the windows at the front of the house and regularly escort Parsnip outdoors into the back garden for fresh air and stimulation under her protective custody. Mummy's Little Tulip tended to remain in the territory at the front of the property so it was unlikely that he would be evident elsewhere. Hilary was also told to actively discourage MLT from the garden if she saw him with a strategically aimed jet of water.

- Hilary was given a grooming and brushing routine to follow on a daily basis to remove dead hair to prevent the accumulation of further furballs and provide a relaxing massage for Parsnip.

- Both Hilary and Duncan were encouraged to play with Parsnip as often as possible. Hilary was also asked to make a catnip pouch that could be given to her as a treat for a short period every other day.

- Synthetic feline facial pheromones were to be used in the kitchen (the plug-in device) to further promote her renewed sense of safety in her own home.

- Various rooms within the house were altered slightly to provide warm beds and high resting places near windows so that Parsnip could enjoy the scenery and have somewhere cosy to sleep.

- I asked Hilary not to reassure or punish Parsnip for the excessive grooming. Punishment would further add to her insecurity and acts of comfort and love could potentially encourage Parsnip to utilize the behaviour in order to seek attention from her owner. I suggested it was best to offer distraction of various sorts including a loud noise or sudden movement.

- Hilary was asked to keep a diary of the grooming she observed, together with an actual measurement of bald patches to assess any changes in size.

The outcome

This particular programme was extended beyond the normal eight weeks because of the medication, since it was important to ensure that any good results were maintained after the drug was withdrawn. After four weeks I received a report from Hilary to say that things were looking extremely promising. She had followed all my instructions and, after a short period of drowsiness, Parsnip had taken to her new feeding challenge and become more active in her playtime with every passing day. She enjoyed her excursions outdoors with Hilary and had even followed the sun round to the front of the house on a number of occasions. Mummy's Little Tulip had actually made a rather fortuitously timed visit to the local cattery for a couple of weeks during the programme and this had further added to Parsnip's sense of confidence. Within four months Parsnip was off the drugs, apparently much happier, and her fur was starting to grow back. Hilary was particularly pleased to note that Parsnip had become more independent and returned to a pattern of behaviour that included much more time in her own company. I think I had worried her about how problematical a dependent relationship can be and she spent quite a bit of time and effort encouraging Parsnip to be more self-reliant.

❖ ❖ ❖

Cats are capable of an even more sinister degree of

'self-mutilation' than over-grooming. Occasionally I will visit a patient who has started to chase his tail. These cats will growl and then pounce on their tails, whirling round in circles and tumbling over themselves to get sufficient hold so that they can bite and chew, sometimes causing horrific injuries. Many cats have had their tails amputated as a result of this behaviour. Sadly, some will then continue to chase their own bottoms instead.

Jazzie – the case of the tail-chasing cat

Jazzie was four years old when I met her and she lived with Julie and Graham. She was a lovely little tabby cat who had been adopted only a few months previously from a local rescue centre. Julie and Graham didn't know anything about her because she had been found as a stray. She settled in very well in her new house and soon got used to the cat flap and coming and going throughout the day. She was extremely affectionate towards me when I visited but I was aware of a slight tension when I stroked her back. Julie mentioned that Jazzie was often quite unpredictable; she would tolerate stroking and cuddles and then suddenly turn and bite for no apparent reason.

One day Julie was watching Jazzie playing in the garden when she noticed a bizarre sequence of behaviour. Jazzie turned to look at her tail and then started to spin round and round and then run off as if trying to escape. Julie thought this was highly amusing until Jazzie actually caught her tail and started to chew it. Since then her tail had become a bloody mess as Jazzie chased and bit it at least once a day. Her owners were distraught because they could only stop her doing it when they were there. If they were out at work they worried constantly about the state poor Jazzie would be in when they got home. The vet had suggested to them that the motivation could be psychological so I was asked to visit to assess the patient for myself.

The facts of the case

Despite the vet's suspicion that this was a behavioural problem it was still essential to keep an open mind. I spent a couple of hours looking for behaviour in Jazzie or information from the history relayed that would point towards an emotional origin to this tail chasing. The significant facts were these:

- Jazzie had an unknown history.
- Jazzie used to bite her owners occasionally when she was stroked.
- She looked a little tense when I stroked her back.
- There were other cats in the neighbourhood; one had tried to get through the cat flap.
- Jazzie responded to steroid treatment.

I always tend to presume that cat stress is a 'cat thing' so I was interested to hear that other cats were around and fairly intimidating. I wasn't, however, convinced that this would be sufficient to induce Jazzie to develop such a bizarre coping strategy for her anxiety. Nothing that Julie or Graham told me indicated that Jazzie was suffering from chronic stress. On the other hand, her behaviour when touched was possibly significant. If she responded to steroids then it was highly likely that her behaviour wasn't psychological; there was probably an itch or pain component associated with the tail chasing. We didn't know Jazzie's background so it was impossible to delve into her history to discover anything relevant. We had to make sure we explored every possibility.

The behaviour programme

- I asked Julie and Graham to monitor the tail chasing in diary form just to confirm that the behaviour diminished after a steroid injection. Environmental changes were noted including weather, presence of other cats, changes in routine, etc.
- If the tail chasing did not respond to the steroids, the vet would prescribe a special 'exclusion' diet to rule out a food allergy. The vet was also considering the prescription of a tricyclic antidepressant.

This was a very short programme because I was suspicious that this was not a psychological problem. During the course of the following eight weeks I was hoping for a further clue to establish a more positive diagnosis. The referring vet was a very astute lady and she was also becoming convinced that something wasn't quite right. One of my suggestions to her, once I had seen Jazzie, was that her behaviour was motivated by pain. She arranged for an X-ray to be taken and found that several vertebrae in her tail had become fused together as a result of a previous trauma. It was agreed that the best course of action was to amputate the tail.

The outcome

Jazzie was a different cat after the surgery. She suddenly decided she loved to be stroked and cuddled and became quite a lap cat. There was no further evidence of discomfort or tail (now bottom) chasing. There had obviously been a source of pain in her tail and her response had been to attack it as a confused reaction to the strange sensation.

There are a number of possible motivations for tail mutilation or chasing. The following is not an exhaustive list because it is such an unusual behaviour that very little is really known about its aetiology.

Localized skin hypersensitivity   This will cause the cat to view the extremity as foreign and therefore behave aggressively towards it. It can result from a flea bite or a food allergy.

Neuralgia   This is a sensory nerve pain caused by trauma or immune response changes.

Displacement activity   This is a behaviour performed 'out of context' because the cat is frustrated in its attempt to perform another activity. It can be adopted as a coping strategy for animals suffering frustration or a sense of conflict.

Feline Hyperaesthesia Syndrome   This is a recently identified condition that describes behaviour including skin twitching and mutilation of the skin on the lumbar spine or tail. The biting action in cats produces a massive sensory response which can be intensified by the angle of the head in relation to the body, hence the theory that the act of tail mutilation may be self-reinforcing if the individual is predisposed. Any skin conditions or anal gland problems are thought to exaggerate the symptoms.

Self-injurious behaviour   Chronic stress can result in the production of unusually high levels of natural opioids. There ensues a complicated addictive relationship between pain and the release of endorphins in the brain, resulting in continuing self-trauma.

Obsessive Compulsive Disorder   This describes repetitive actions that, like displacement activity, occur out of context and potentially interfere with normal function. These behaviours can start as a result of unresolved stress and continue because they are neurochemically rewarding or occasionally positively reinforced by the owner's attention.

As you can see this is a complicated problem and best dealt with by veterinary surgeons who are also behaviour specialists.

Pica

Pica is a term used to describe the consumption of non-nutritious materials and it describes a habit indulged in by a very small percentage of the domestic cat population. The substances chosen are many and varied but the most delicious appear to be fabric of varying descriptions, cardboard, rubber, polythene, carpet and plastic. Some cats are perfectly content to lick or chew rather than consume but the behaviour can still be rather disturbing.

An extreme example of this behaviour is referred to as 'wool eating' and is relatively common in Siamese and other derivative breeds. Although wool is usually consumed initially the habit often generalizes to other natural fabrics. Behaviourists believe there is a genetic element to the problem and that, in some way, the susceptible cat's brain works slightly differently from others. One theory suggests that the act of chewing causes 'happy chemicals' to be released in the brain, giving the recipient a feeling of intense pleasure. If you ever see a cat 'wool eating' you will notice how he moves the material back in his mouth to his carnassial (shearing) teeth and then adopts an expression of sheer ecstasy as he chews.

The treatment for this problem is often fraught with difficulties as 'wool eating' appears to be highly addictive. Some young cats grow out of the problem but others continue with stubborn determination throughout their life. Owners

have to adapt their home accordingly and all materials are kept well out of reach. This may sound easy in principle but, in reality, it is extremely difficult, particularly as the cat is so motivated to chew. Many owners resort to providing small amounts of chopped-up wool to satisfy the craving and hope that it passes through the cat's digestive system without blocking it.

I dread seeing these cases because they are so hard to resolve. I have had good results in the past by using techniques to stimulate the cat in other ways. I recommend plenty of play, food foraging and the provision of the occasional cooked knuckle bone to give the cat something safer to chew. The referring vet often prescribes a long course of a drug of the same type as Prozac to try to balance the brain chemistry and eliminate the urge to eat wool once and for all. Some resolve whilst others remain 'wool eaters' for ever.

Pica can also occasionally be motivated by other rewards. I once saw a pair of Russian Blue cats who persistently ate bars of soap. Since they had their poor owner firmly wrapped around their paws, it was no surprise to find that their motive was purely to seek attention.

If you have a cat who licks, chews or consumes something he shouldn't, it is important you seek help as soon as possible.

The sorts of items eaten can cause intestinal blockages that require surgery and chewing electric cables (another favourite with Burmese) is extremely dangerous, so it is not a problem to be taken lightly. Your veterinary surgeon can help you in the first instance and if necessary refer you to someone like myself.

Night-time vocalization

Night-time vocalization is a surprisingly common habit of the elderly cat. I conducted a survey back in 1995 into the behaviour of the geriatric cat and a total of 1,236 owners completed detailed questionnaires. Twenty-eight per cent of those surveyed said that their old cats called for attention at night and stopped only when they received attention or reassurance. Of these 346 cats, 54 per cent started the behaviour between the ages of ten and fifteen. As a cat's ability to protect himself declines there appears to be a higher dependency on the owner for security. Having enjoyed additional attention during the day, these cats often feel in need of reassurance when their owners are not around in the night. Having tried successfully a number of times to elicit a response from their owners (the harsh distressed yowl is difficult to ignore) they continue to perform the ritual as a habit. A number of owners from the survey reported that the calling stopped when the cat was allowed to sleep in the bedroom. However, often these cats jump off the bed and wander off downstairs only to repeat the crying all over again.

There are also a number of physical reasons why this calling may occur. It is possible that a chronic deficiency in the oxygen supply to the brain could produce symptoms of senility and short-term memory problems causing confusion at night, especially if cognitive dysfunction has triggered a change in the sleep/wake cycle resulting in wakefulness at times when the cat had previously been asleep. High blood pressure causing general discomfort, headache and disorientation could also promote this sort of distress response.

Night-time vocalization is often reported as one of the behavioural symptoms in cats suffering from hyperthyroidism, a condition seen frequently in the elderly cat. A tumour on the thyroid gland causes metabolic changes including increased heart and respiration rates, increased appetite and weight loss. I have had three cats myself with this condition, all of whom cried pitifully at night.

Dottie – the case of the crying cat

Dottie was a sweet little black and white moggy who lived with Isobel and her family. She had been born before Isobel's two children and she had seen both of them grow up during her eighteen years. She looked amazingly robust for her age although she didn't really do much. She still used the cat flap to go outside but much preferred the more dignified exit on request through the back door. Isobel kept Dottie in at night and she always used to sleep peacefully on the master bed.

Since the death the previous year of her best friend Jack, the family's terrier, she had not been herself at all. She had started to be restless at night and wander around the house yowling in the most distressing way. Isobel would rush to her side and comfort her; she would even occasionally walk around the house rocking Dottie gently in her arms to try to get her back to sleep. The problem continued and, every night, Dottie would call out at four o'clock in the morning and Isobel would rush to her, reassure her and stay with her until it was time to start the day. Isobel was shattered and completely at her wits' end. She had tried to ignore Dottie, on advice from her vet, but she only remained steadfast for two nights. Dottie cried even louder until Isobel relented and answered her call.

When Isobel contacted her vet about the problem he was duly diligent and performed a general health check and checked Dottie's thyroid function. She appeared perfectly healthy for an eighteen-year-old cat so he referred her to me.

The facts of the case

I chatted with Isobel for a while and observed Dottie, who was enjoying all the extra attention. The following points were relevant and helped me make my decision about the programme we needed.

- Dottie was eighteen years old (about eighty-eight in human terms).
- The behaviour had started after the death of her friend Jack.
- Isobel went to her every night because she was worried that Dottie was distressed.
- When Isobel ignored Dottie for a couple of days the behaviour became worse.
- Dottie had been given a clean bill of health by her vet.

In the absence of a medical cause for this problem we were probably looking at an age-related insecurity. Old cats crave routine and a predictable lifestyle and Jack's death was extremely distressing for Dottie. She started to feel insecure and vulnerable at night when the house was quiet, and she became distressed, hence the harsh call. She may even have been disorientated as a result of her advancing years and general decline in mental agility. Isobel came running in response to her call and all was well. Many cats cannot settle at night when they get older and the nocturnal wandering was confusing poor Dottie. She then continued to call every night when she felt she wanted company from her devoted owner.

The behaviour programme

Dottie needed some predictability and routine back in her life. She also needed a bit more encouragement to be active during the day, both mentally and physically. It's very true for cats and humans that 'if you don't use it, you lose it'! Dottie needed to try to increase her sense of self-confidence by getting her brain active again.

- Dottie needed a 'bedroom' where she could go at night and feel safe. Isobel provided a warm bed by the radiator, food, water and a litter tray, and established the kitchen as her safe haven at night.
- Every evening Isobel gave Dottie a small bowl of warm cooked chicken and wished her 'good night'.
- Any calling at night was then ignored totally, even if it got worse (which would merely be Dottie getting frustrated and trying harder to get a response).
- Isobel and the children were asked to play gentle games with Dottie using a length of string attached to a stick and various home-made cat toys.
- Dottie should be given love and affection during the day on demand.
- Dottie also loved being groomed so we incorporated this into the new daily routine.

The outcome

From the first night the calling reduced significantly. Dottie wasn't wandering around and the room felt comfortable and safe. She enjoyed the games during the day and the family reported there was a slight spring in her step. She loved the chicken at night and even started to put herself to bed at the allotted time. Dottie lived happily for another two years with Isobel and her family until she was peacefully put to sleep shortly after her twentieth birthday suffering from kidney failure.

Feline Idiopathic Cystitis

There are many other medical reasons why your cat may display unusual or worrying behaviour. Feline Idiopathic Cystitis is becoming increasingly prevalent, the first sign of which is often house soiling (see Chapter 4). Whilst we associate cystitis with frequent squatting and passing small amounts of blood-stained urine this is not always the case with a more chronic problem. If your cat should start to urinate inappropriately in the house it is always worth a visit to your vet for a thorough examination.

Seizures

Some seizure activity is not quite as dramatic as an epileptic fit but the behaviour can be equally alarming. I have seen cats that appear to 'moon walk' as a result of mild seizures or even attack their owners. Tumours on the brain can also result in behavioural changes although they are, thankfully, comparatively rare.

Arthritis

One of the questions in the elderly cat survey I conducted in 1995 asked the owners if their cats were suffering from any illness or disease. The most common complaint cited by the owners surveyed was arthritis. This degeneration of the joints, often exacerbated by previous fractures, can be very debilitating. Your cat may walk differently, go outdoors less (they just can't jump that fence any more) and generally condense a great deal of their normal range of activity. They may also start to become less tolerant of cuddles and stroking and a warning hiss, if unheeded, may even lead to an aggressive attack. This change in behaviour would undoubtedly be related to pain and a veterinary examination in combination with X-rays could well reveal the problem.

CHAPTER 10

Bonding Problems

MOST OF US ARE PERFECTLY HAPPY TO ACCEPT THAT WE ARE AT the beck and call of our feline friends. We seem to relish the way our cats control our every movement in order to obtain the maximum pleasure. We start with every good intention to feed them veterinary formulated quality cat food and end up supplementing this balanced diet with crevettes, honey roast ham and corn-fed chicken. Why? Because our cats want it. This always seems an adequate excuse for just about any indulgence. But is it really they who demand this haute cuisine? Is it really they who require the specially designed four-poster bed or the diamond-studded collar? Cats must live in a constant state of amazement at every little gesture that the average owner makes in the name of love. They are great opportunists, so why wouldn't Tigger hang round the fridge if something tasty came out of it every time he gave a pitiful miaow?

Spoiling a cat generally does no harm whatsoever providing you give the culinary extras in moderation. Unfortunately some of us believe the maxim 'I love therefore I feed' and feline obesity is a growing problem. Some may believe that the size of their cat's belly is in direct proportion to the amount of love they receive but I beg to differ. Obesity in cats can cause diabetes, heart disease and joint problems and potentially shorten the lives of these beloved pets. There are other ways to love.

Occasionally the extent of the bond between cat and owner reaches a level that merits the term 'dysfunctional'. The dilemma implicit in dealing with such an issue in a DIY manual is that the last person to see the problem is the over-attached owner herself. Yes, in my experience these owners tend to be female. As women we all have a need to nurture and care; occasionally that need is not fulfilled in our relationships with other humans. Cats have an ability to provide what is perceived to be unconditional love and a relationship with them just seems so much easier, somehow.

I have often described an over-attachment as 'an emotional bond with a pet that is so intense that it is detrimental to the physical or psychological well-being of either the human or the animal'.

Most cases of problem behaviour that have developed as a result of over-attachment involve anxiety-related indoor soiling, attention-seeking urine spraying (or other unsavoury activity) and aggression. The most common presentation is unusual cat/owner responses on both sides. For example, a slightly incompetent and nervous cat goes to live with a caring, solicitous, emotional owner and the eventual result can be a 'learned helplessness' in the cat and an over-attached relation-ship. The owner is so busy reassuring the cat that it will be suitably protected that the poor shivering wreck sees danger everywhere and collapses in a heap, completely unable to do anything unless the owner is present. If the owner should be absent for any length of time the cat will stop eating and prob-ably wet the duvet. The other scenario tends to feature a highly intelligent, sensitive cat (for example a delightful but high-maintenance Siamese or Burmese) meeting the same caring, solicitous, emotional owner. The outcome here is very different, as undesirable attention-seeking behaviour, such as urine spraying or cable chewing, often becomes the problem.

This chapter may fall on deaf ears, but if you feel that your home is no longer your own and you want your life back then you may have slightly overdone the whole compliant owner thing.

Is this you?

Do you have sleepless nights because, somehow, your cat takes up more of the bed than you do?

Do you stop what you're doing and attend to your cat's needs any time of day or night because he miaows?

Do you carry your cat around everywhere like a hairy brooch because he likes it but end up with a repetitive strain injury in your shoulder?

Do you read the newspaper with a cat's bottom firmly planted on the financial section?

Do you get up at 3 a.m. to microwave fish because that's when your cat wants it?

Do you constantly replace your telephone cable because

your cat chews through it when you are on the phone?

Do you avoid going on holiday because the one and only time you did your cat went off his food and you had to get an early flight home?

Never forget that the pleasures of pet ownership should be a two-way thing. This could mean that you need to allow your cat outside sometimes, even though you worry about him, so that he can run around and do what cats do naturally. It also means that you have every right to a good night's sleep and a cosy bed for your cat in the kitchen is not hell on earth.

If you want to ensure your relationship remains on an even keel then here are a few tips.

- Start as you mean to go on. There are many veterinary formulated foods that will provide your cat with everything he or she needs. The occasional tidbit is fine, but it is important to monitor your cat's weight because obesity can creep up slowly. You should always be able to feel your cat's ribs if you run your hand down his side.
- It is perfectly all right to reject your cat's attentions when you are busy elsewhere; he will not love you any less. It is important for him to learn that you don't always get what you want when you want it.
- If your cat should demand attention at night do your best to ignore him. If you don't, you will end up with dreadful sleep deprivation symptoms and a cat that is very difficult to live with.
- Try to ensure that your cat has other interests apart from you, including visits outdoors whenever possible and activities around the house. Whilst a loving relationship is a joy it shouldn't be the only focus in your cat's life.

Here are a couple of cases that you may find helpful if you are suspicious that your cat has the upper paw in your relationship.

Murphy – the case of the attention-seeking cat

Murphy and Marco lived as indoor cats with Jane and Trevor in a spacious flat in central London. Murphy was six years old when I met him. Jane and Trevor acquired him when he was just seven weeks old. A friend had found a stray cat and kittens and Murphy was one of the litter of five that needed good homes. They chose him because he was a pretty tabby but they realized they had a difficult job on their hands because he was extremely timid when they first saw him. When Trevor picked up the little kitten he pushed his head into Trevor's arm and this was so endearing they had to have him. Trevor and Jane lavished love and attention on Murphy and he grew into a beautiful healthy adult cat. He did, however, remain clingy and dependent on his owners and went to Trevor in particular if anything bothered him. Visitors, loud noises, changes in routine and sudden movements all sent Murphy rushing to the side of his owner.

Jane and Trevor were very concerned about Murphy's apparent dependency so they decided to adopt a kitten to keep him company. They chose a black kitten that had been reared by a local cat charity, and called him Marco. He was confident and sociable and a complete contrast to Murphy. As is often the case, Murphy was not exactly thrilled when Marco came along but he tolerated him reasonably well. Marco developed into an independent and laid-back adult. It was perfectly clear that the time he spent with his owners was for pleasure on his own terms. There was no way he needed them, but if there was any competition between Marco and Murphy for access to an owner's lap the younger cat would win every time.

Both cats used to enjoy the company of their owners at night as they curled up on the bottom of their bed. A couple of years prior to my visit Murphy had started waking Trevor up in the early hours of the morning by clawing at his feet and crying. Trevor would dutifully rise, go to the kitchen to feed Murphy, and then return to bed. Murphy wasn't having that and pestered Trevor until he remained in the living room with the cat until it was time to get ready for work, just to keep him quiet. This was fine during the winter months, since Murphy started his crepuscular howling at a reasonable hour, but dawn breaks very early in the summer and Jane and Trevor started to feel in desperate need of a decent night's sleep. They eventually decided they had no choice but to exclude the cats from the bedroom at night to try to resolve the problem. This was a shame for Marco who remained uninvolved in the disturbances but it had to be 'one out, all out'. The following nights were a disaster as Murphy exhibited excellent problem-solving ability by learning how to hurl himself at the door handle to gain entry at the appropriate time. Trevor continued to get up and feed him but every evening, before retiring, he would try various methods to secure the bedroom door. Eventually, with the aid of wire, string, cardboard and a heavy chair, Trevor and Jane managed to turn their bedroom into an impenetrable fortress. Murphy was not deterred. His new ploy involved sitting directly in front of the bedroom door with his furry face pressed up against it like an outside broadcaster's microphone, and screaming and screaming until Trevor came, fed him and played with him until it was time for work.

Over a period of time both Trevor and Jane realized that the situation was getting ridiculous. They were both irritable and showing signs of confusion and short-term memory loss due to the lack of sleep. Trevor had started working at home a couple of days a week but this was proving impossible because he had to do everything with Murphy glued to his lap or shoulder. Something had to change and it had to happen quickly before Jane and Trevor went mad.

The facts of the case

It's fascinating watching clients and the way they interact with their cats. Most relationships involve communication between owner and cat that the former carries out without conscious thought. As I was discussing the problem with Jane and Trevor they were both touching and acknowledging Murphy constantly. He in turn was moving backwards and forwards between his owners in response to their actions. When I asked them if they were aware of their constant attention towards Murphy they seemed puzzled. They suddenly became self-conscious and, sure enough, they could see what they were doing. They focused constantly on Murphy and this had become so habitual they were not even aware of it.

These are some of the most relevant facts gleaned during the consultation:

- Murphy was insecure from a young age and sought re-assurance from his owners.
- He was kept exclusively indoors.
- Jane and Trevor provided constant attention.
- There was an uneasy alliance between Murphy and Marco, with the latter often denying Murphy access to his owners.
- Trevor was getting up early in the morning in response to Murphy's actions.

- Murphy tried even harder to seek Trevor's attention when they banned him from the bedroom at night.
- The situation appeared to be getting worse.

Some kittens are born timid and, in the absence of important early socialization, they can remain so when they become adults. Many such cats will shun human attention but occasionally a nervous kitten will seek solace in the arms of a caring owner. Murphy was one such kitten. Trevor in particular had lavished attention on Murphy from the very first day and this had created a relationship born of dependency. As Trevor reassured Murphy in the face of any situation, threatening or otherwise, he taught the cat to rely on him for comfort and safety. Murphy also learned very quickly that any request for attention was always fulfilled. Cats kept exclusively indoors rarely expend the energy they would if they lived a more natural outdoor lifestyle. Cats are nocturnal hunters with particular preferences for dawn and dusk so it would be perfectly natural for Murphy to be active at those times. Trevor's provision of food was all very well but Murphy probably just wanted to be entertained and stimulated. Forget the food; let's have a bit of company!

Murphy's world revolved round Trevor and all his obvious charms as the perfect provider. He was a source of food, play, comfort and company and a good all-purpose entertainer at any time of day or night. Whilst this sounds like the perfect relationship (cat loves owner, owner loves cat) it is clear that it is potentially a nightmare. Pet ownership is admittedly all about pleasure but it has to work for both parties. Trevor was not getting pleasure and, ironically, neither was Murphy. He was demanding more and more attention and he had become so focused on his owner that he was finding it hard to function on his own. Marco was causing problems for him also as he would on occasion prevent Murphy from approaching his owners, just because he could. This had added further stress and created even more tension between the two cats. We had to break the cycle for everyone's sake.

The behaviour programme

I explained this to Trevor and Jane and they agreed that things had to change. I made the following recommendations:

- Trevor and Jane had to start to interact differently with Murphy. They had strict instructions to ignore Murphy when he started his night-time howling. Ignoring would involve no verbal communication and no action that could be interpreted as a response. They had to stay firmly fixed to their bed and not venture forth until the alarm went at a reasonable hour. I explained that it was almost inevitable that Murphy would try harder to get their attention and it was imperative that they did not relent.
- Trevor had to understand that denying Murphy's requests for food or attention was not an act of cruelty. He would still be provided with attention but it would be rationed at times to suit Trevor and Jane.
- When Trevor was working at home it was also important for Murphy to understand that he couldn't have unlimited access to his owner. Trevor was asked to sit near to his desk to prevent Murphy from jumping onto his lap. He was also asked to ignore him with no eye contact or verbal communication. If Murphy jumped up and became too disruptive he would have to be shut into the bedroom or bathroom (with litter facilities, food and water) for a period of time to allow Trevor to finish his work.

- It is important in these cases not to leave the cat in a void. We had to ensure that Murphy had a choice of alternative activities to provide a suitable distraction from Trevor. Murphy and Marco's environment within the flat represented their entire world and it wasn't the most stimulating or dynamic space for cats. We set about making it a veritable adventure playground. Shelving was erected to provide both cats with opportunities to rest and observe life from a higher level.
- Trevor was a real DIY enthusiast so we agreed that he would construct some wooden climbing frames that could be placed against the wall to give access to the high shelving. They would also make great scratching posts.
- Trevor also put a rectangular section of carpet on the wall for climbing and scratching.
- Murphy and Marco were eating a dry diet so it was agreed that they would start to forage for their food. Trevor and Jane were asked to hide the daily ration in various locations and allow the cats to find it. When two or more cats are in the household it is important to monitor this new feeding regime because some cats can be more adept at finding the food than others.
- Extra water bowls were placed around the flat and I recommended that Trevor and Jane purchase an indoor water fountain or pet drinking fountain as a source of running water. Murphy used to jump up onto the basin in the bathroom and scream to summon Trevor to turn the tap on. This really had to stop too but I asked Trevor to ensure that Murphy was using the fountain or other sources of water first before he ignored Murphy's requests. It is extremely important that cats drink plenty of water when they are eating a dry diet.

day, they spent quality time with Murphy and Marco. The two cats even seemed to be getting on better and both Jane and Trevor felt that the sense of tension and competition between them had become less evident. The extra work involved in providing stimulation for Murphy outside the relationship with Trevor paid off and he continues to behave well. He's still active at dawn but he uses up his energy by running up the carpeted wall and chasing ping-pong balls instead of bothering his owner.

Peanut – the case of the 'home alone' cat

Peanut was a beautiful young Birman with sapphire-blue eyes. When I met him and his owner, Fran, he was approaching his second birthday. Peanut and Fran lived in a house in Surrey. Fran was a busy lady who desperately wanted company after long hours away from home. She had decided to buy two kittens (so that they weren't lonely when she was out) and eventually decided on Peanut and his brother, Cashew. Tragically Cashew died when he was only a few months old of a dreadful illness called Feline Infectious Peritonitis. Both Fran and Peanut were inconsolable and the next few months were spent in each other's embrace to try to soften the blow of their bereavement. Time is a great healer and they both came to terms with their loss. However, the bond that had formed between the two had become extremely strong and Peanut would cling to Fran every minute she was at home. He would go to the bathroom with her, he would sleep with her, he would be under her feet (or halfway up her leg) when she was cooking, and generally would perform the role of a small furry shadow. Fran loved the relationship and looked forward to coming home every evening to such an affectionate companion.

Peanut was kept exclusively indoors on the recommendation of his breeder. Fran left a box of toys in the living room for his entertainment but she found that he had got into the routine of sleeping during the day until she returned from work. About eighteen months after Cashew's death, Fran decided it was about time she had a holiday. She felt that Peanut had recovered sufficiently from his brother's death and would cope with her being away for a week. She arranged her sun-and-sea break and asked a neighbour to visit twice a day to feed and entertain Peanut.

Fran had a great time but her pleasure was short-lived when she returned to her beloved Peanut. She noticed he seemed rather strange when she got home; a little distant and un-settled. She thought he was probably sulking because she had left him for a week but she certainly wasn't prepared for the surprise gift he had lovingly arranged on her duvet. Her beautiful Egyptian cotton bedding was festooned with faeces in the most disgusting 'dirty protest' that Fran could possibly imagine. How could he? What a dreadful punishment for a week away. Fran could barely look Peanut in the eye as she went straight round to her neighbour to see if she could shed any light on this unpleasantness. Fran's neighbour was dis-traught; she had not entered Fran's bedroom and just presumed that Peanut was a little constipated given the lack of deposits in the litter tray. Peanut had apparently been very vocal when she came every day to feed him but she wasn't aware he was then disappearing upstairs to perform his dirty deed.

The facts of the case

I listened to this story with great interest whilst watching the interaction between Peanut and his owner. He watched her and touched her and she watched him and touched him . . . constantly. This really was an intense relationship. Here are a few more relevant points.

- Fran and Peanut's relationship had intensified after the death of Cashew.
- Fran's week away was the first time that Peanut had been left home alone since the relationship had started.
- Peanut didn't have much stimulation in the home apart from his interaction with Fran.
- Peanut lived exclusively indoors.

The death of a companion cat is distressing for the whole family and Fran would have gained a great deal of comfort from the idea that both she and Peanut were grieving. In reality she was probably merely creating a dependent relationship as Peanut turned to her for all his entertainment and security. Indoor cats need a great deal of stimulation to compensate for their lack of the exciting challenge of a natural outdoor life. Peanut wasn't really getting this sort of activity so his entire focus became interaction with Fran or waiting to interact with Fran. When she went away Peanut would have suddenly found himself alone. Deprived of his one source of entertainment and security he would have had a strong sense that the defence of his fortress lay solely in his paws. It would have felt like an enormous responsibility and poor Peanut probably panicked. He may well have passed faeces on Fran's bed in a marking gesture (see Chapter 5) to signal to all invading forces that he was a formidable opponent. Combining his own scent with that of his owner would have reassured him considerably.

The behaviour programme

I explained to Fran that we needed to encourage Peanut to explore life outside their relationship. It wasn't healthy for either of them (or the bed linen) to continue with this level of dependency. We discussed the following programme.

- Fran was asked to interact with him differently and avoid all the touching and focusing that seemed to be such a large part of their time together.
- The few toys in the basket were not sufficient incentive for Peanut to entertain himself. They didn't move and they weren't particularly exciting. Fran was encouraged to make new toys including fishing-rod games and home-made catnip mice (see Chapter 11 for more information about exciting toys).
- Peanut was encouraged to forage for his food (this really is a useful technique) and his dry food was placed inside egg boxes and toilet roll tubes to challenge him when he was hungry.
- Peanut used to shout at Fran to turn the tap on when she was in the kitchen so that he could play with the running water, so she bought an indoor water feature for him to use as a drinking bowl.
- Fran attached a section of carpet to her wall next to a shelving unit so that Peanut could run up the wall and sit on the high shelves.
- I also suggested that Fran did a little research to find a suitable local cattery for future holidays. Some cats are better off away from home while the owner is absent, precisely to prevent this type of panic response.

The outcome

Basically, Fran enriched Peanut's environment and gave him loads of things to do whilst withdrawing gradually from the relationship. Peanut continued to be pleased to see his owner when she came home from work but he soon had plenty of other things to occupy his mind. Fran found a lovely small cattery a few miles from her home and she booked well in advance for her next holiday.

CHAPTER 11

The Indoor Cat

I CANNOT POSSIBLY COMPLETE A BOOK ABOUT BEHAVIOUR problems in cats without devoting some time and attention to those that are housebound. Ten per cent of cats in the UK are believed to live exclusively indoors and this number is increasing. Whilst this implies that the majority of cats are roaming free, the reality is very different. Many cats actually have their access outdoors restricted to some extent. Some are timid and find the prospect of mixing with more assertive individuals at the bottom of the garden rather daunting. Elderly cats often err on the side of caution when contemplating an encounter with next door's tomcat and choose to remain in a warm and secure environment. Cat welfare charities recommend that cats are kept indoors at night to protect them from any number of hazards more evident during the hours of darkness. We all love our cats; sometimes the perils of outdoor life seem too much to bear and we consciously decide to become our pets' bodyguard, only allowing them outside under strict supervision. However did they manage as a species before we got involved?

I will not be drawn into a heated 'should we, shouldn't we?' debate about confining cats indoors. In an ideal world no cat would have restrictions on its movements and no cat would ever be bred that was so modified it had to be confined for its own safety. But, as we all know, this is not an ideal world and I fully appreciate that the benefits of ownership might appear to outweigh the withdrawal of a cat's freedom. My argument would merely state that ownership of a cat kept totally or partially confined indoors requires extra effort to ensure that the cat is suitably compensated.

Cats that have free access to outdoors will potentially enjoy a full and natural lifestyle. The constantly changing outside world will provide all the mental and physical stimulation required to keep the individual healthy. However, as soon as the cat's world becomes shrunk to a static area within four walls it is difficult to see how he can entertain himself to the same extent. The modern trend towards minimalistic living must be the ultimate tragedy for the indoor cat. Whilst the clean lines of single pieces of furniture against stark walls may appear aesthetically pleasing to our taste I often wonder what a cat must make of it all. Where are the nooks and crannies? Where are the secret areas to explore? Where are the high resting places? Where are the moving objects?

If the cat becomes bored, frustrated, depressed or generally stressed in such an environment it is hardly surprising that things start to go wrong. As I so often say, 'The devil makes work for idle paws.' Behavioural problems that are caused,

in part or wholly, by an inappropriate environment include
- inter-cat aggression
- inappropriate urination or defecation
- urine spraying
- over-grooming, self-mutilation, fur plucking
- stereotypies (compulsive disorders)
- obesity
- anxiety, depression
- over-attachment
- attention-seeking
- aggression to humans
- Feline Idiopathic Cystitis

It is worth noting that all these potential problems could be prevented if we took a little time and trouble to adjust our homes to suit our cats' needs as well as our own. It isn't necessary to consider these changes as chore-based, rather negative elements of cat care. They can be fun and greatly enhance your relationship with your pet. Sometimes your cat would appreciate your showing your love in other ways apart from stroking and petting.

What is environmental enrichment?

Environmental enrichment is a phrase that is becoming very popular in the veterinary world. It embraces an animal's natural habits and behaviour and endeavours to make provisions that stimulate and challenge the individual and enable it to perform natural behaviour in an artificial environment. This works well in modern zoo enclosures and is no less relevant for Sooty in his laminate-floored lounge.

Important resources for your cat

Resources within the home represent all those things that provide nourishment, entertainment, stimulation and security for the pet cat. They may be items purchased or made specifically for your pet or objects of furniture utilized by the family in general. If you ensure there are sufficient quantities and types of resources for the number of cats in the household it means that every effort is being made to promote their well-being and emotional health. These resources include

- food
- water
- vegetation e.g. a source of grass
- litter trays
- social contact
- high resting places
- private areas
- beds
- scratching posts
- scent stimulation e.g. catnip and valerian
- predatory play
- toys
- novel items
- fresh air

I will address each one of these resources individually and as you work through the list you may find a number of suggestions that could add hours of fun and entertainment to your cat's typical day indoors. Many owners spend a fortune on cat toys every year only to find that most are dismissed as total nonsense by their discerning pets. I love making my own toys on the basis that most cats would rather play with rubbish anyway! Some of the essential provisions discussed, such as food, are often taken for granted but even these can be offered in such a way that the cat is stimulated and entertained.

Food

A natural diet would require hunting for several hours a day and the consumption of as many as a dozen small rodents. How many cats living naturally eat twice a day from a predictable source? Cats spend up to six hours a day hunting, foraging, stalking, catching and consuming prey. The availability of food twice a day, or even 'ad lib', in a food bowl in the kitchen, does not represent any kind of challenge whatsoever. The normal feeding regime for the average pet cat potentially leaves a void of five hours and fifty minutes that it needs to fill with other activities. The time is often filled with sleep in an otherwise static and uninteresting environment. Exciting and stimulating challenges utilizing moist or wet food is going to be difficult since I wouldn't suggest for one moment that you secrete a bowl of tinned food in the bottom of your wardrobe for your cat to 'hunt'. The possibilities, however, are endless if you are feeding a dry preparation. The idea of 'food foraging' works on the principle that obtaining smaller amounts of food more frequently in a variety of locations represents a more natural way of feeding for a top of the food chain predator like Tiddles, Felix and co.

It would not be unreasonable to expect our cats to work a little harder for the food they obtain throughout the day. After all, it wouldn't come easily in the natural life of a predator. They should be able to obtain food in locations throughout the house, both on high and ground level. When you first endeavour to secrete the biscuits in various hiding places your cat will probably follow you around and gobble up the stash in the usual five minutes. This is not the object of the game despite its being another example of how incredibly opportunistic cats are at conserving energy. It is always difficult to be one step ahead of even the average cat but, in this instance, it may be necessary to shut your cat away or have airtight containers in various locations rather than one place where food is always stored. This will enable you to secrete the biscuits randomly when your cat least expects it.

When you first start this regime it will be very tempting to abandon the concept of food foraging as it must surely constitute cruelty of the most sadistic kind. Your cat will sit howling by the now empty food bowl in the kitchen and stare at you with pitiful eyes and sunken cheeks (obviously this trauma has caused immense weight loss). Some cats are less capable than others of dealing with change in a predictable and routine existence and this dramatic turn of events could cause some stress in the more delicate emotional types. This doesn't mean that the foraging should necessarily be abandoned since it could well promote increased self-confidence as your cat faces new challenges. The transition with these cats should be gradual and include a small amount of food still available in its usual place. Other locations should initially be close to the original and moved further away over a period of seven days. This should ease the burden on even the most sensitive individual. Once your cat is used to obtaining food in novel locations the acquisition can become more challenging:

- Build cardboard pyramids of toilet roll or kitchen roll tubes. Place five tubes side by side and glue together. Add four tubes on top and stick them together in a row and to the tubes beneath. Continue to build with decreasing numbers in each row (5, 4, 3, 2, 1) to form a three-dimensional triangle. Place single pellets halfway along each tube and allow the cat to obtain the food by using its paw. Variety can be introduced by leaving some tubes empty and placing five or six biscuits in others. If your cat really gets to grips with this idea you will find him sitting staring at an empty food triangle awaiting the appearance of the 'mouse' with quivering anticipation. Great fun. These triangles may need replacing regularly so keep an ongoing stock of toilet roll tubes. If your cat is blessed with unusually large paws you may be advised to use the tubes inside wrapping paper instead for a much larger circumference.

- Place biscuits inside small cardboard boxes with the lids slightly open to encourage the cat to knock the box over or remove the food with its paw.
- Place biscuits inside cardboard egg boxes with the lid partially shut.
- Place biscuits inside paper bags with the top lightly folded shut.

- Place a couple of biscuits inside a rolled-up piece of paper and throw it for your cat. He will pounce on the 'prey' and pull it apart to reveal the prize.
- Stick two yogurt pots together to form a diamond shape. Make holes in the pots, approximately the size of a two-pence coin, using a soldering iron (this will ensure the edges are not sharp). Attach a piece of elastic through a smaller hole in the top and hang about two or three feet off the ground. Place dry food inside and encourage the cat to tap and agitate the pots to obtain the food as it falls through the holes. To guarantee your cat's safety make sure that the elastic has a breakaway section just in case he gets caught up in it during a frenzied attempt to get at the biscuits.
- Some cats will enjoy chasing biscuits if you throw them across the floor. This is a nice opportunity to have some interaction during a feeding session.

Water

If you are utilizing the dry-diet food-foraging approach it is essential that there is every opportunity for your cat to drink. Whilst most commercial wet foods contain 85 per cent moisture, the dry formulations require extra drinking to maintain a good hydration balance and urinary tract health. The majority of owners (me included until a few years ago) always provide water in the same location as the food bowl. We like to have a glass of water with our meal so why wouldn't Tigger? It doesn't work that way for cats, who naturally hunt for food and search for water on separate occasions to satisfy either hunger or thirst. The presence of water near the food can actually deter some cats from drinking sufficient fluid and this could be dangerous on a dry diet. Finding water elsewhere can be extremely rewarding; how many times has your cat drunk from the glass by your bedside table? There should be at least 'one water container per cat in the household plus one' in various locations completely away from the food. Some cats object to the chemical smell from tap water so filtered or boiled water can be used. There are various ways to provide water:

- Pet water fountains. There are an increasing number of these available, some of which filter the water.
- Feng shui water features. Whilst these are not designed specifically for cats they can be just as entertaining as the genuine pet fountains if you stick to the simple 'water running over pebbles' type.
- Tumblers. As many cats like to steal from our glasses why not provide them with one of their own? Heavy resin tumblers are now available that are non-breakable and these will be safer than glass.
- Ceramic, plastic or stainless-steel bowls. There will still be a place for these in your home to ensure there is plenty of choice. If you have a large bowl available it can be fun to float a ping-pong ball in the water. This attracts the cat to play with the ball and the movement of the water may encourage him to drink.

Vegetation

A source of grass is essential for the house cat to act as a natural emetic to rid him of any nasty furballs. This can be purchased as commercially available 'kitty grass', or pots of grass and herbs can be grown indoors specifically for this purpose. The Feline Advisory Bureau produces a comprehensive list of plants and flowers that are potentially dangerous for cats. Despite the provision of grass specifically for them, many cats will still chew house plants so reference to this list will prevent you from buying anything nasty.

Litter trays

When cats are confined we take all responsibility for the appropriate provisions. The most difficult decision we make on their behalf is that about toilet facilities. Elimination is normally a very private thing and outside cats will have very specific criteria for the sort of location that represents a suitable loo. It is almost impossible to understand what each personal preference would be but we can certainly reduce the odds of getting it wrong by following some simple rules.

- The locations should be discreet and away from busy thoroughfares.
- Trays should be located well away from feeding areas or water bowls.
- The trays should be cleaned regularly.
- The litter substrate should reflect the cat's natural desire to eliminate in a sand-like substance (remember, they are all African Wild Cats at heart).
- Never expect an indoor cat to share a tray with another.

The ideal number of litter trays in an indoor environment is 'one tray per cat plus one', placed in different discreet locations. They can be covered trays with hoods or open shallow containers but it is important that the areas represent a place of safety where the individual does not feel vulnerable. Some assertive cats in multi-cat households will sit on top of covered trays or stand in front of the opening to intimidate the less confident individual. A variety of trays, open and covered, would therefore be advisable in multi-cat households to avoid any risk of house soiling because a cat is too scared to use the normal facilities.

Social contact

Whatever provisions you make in your home there is no substitute for some exciting and loving social contact. It is important to allow your cat to dictate the type and pace of this contact so try to tune in to their needs as individuals with different characters. It is best to respond to a cat's approach rather than chasing it round to initiate contact. This persistence can be irritating or, at the worst, distressing for your cat. Predatory play, grooming and verbal communication is important social contact between owner and cat so these areas should not be neglected in favour of the more popular hugging, squeezing and stroking. Some cats (despite my insistence that most don't) enjoy the company of their own species so the introduction of two initially may be useful if you are away from home during the day. There is always a risk that they will have problems as they get older but the provision of the right number of resources within the home will guard against the need to compete. Some cats will stare at fish tanks, watch your child's gerbils for hours on end or tease the family dog mercilessly so it is important to remember that company can come in different forms.

High resting places

Cats are natural climbers and it is important that the home environment reflects this by providing opportunities to rest and observe proceedings from an elevated vantage point. This will encourage essential exercise and is particularly important in a single-storey home without stairs. Cats like to go up when they feel threatened so if you don't have an upper floor your cat will need to find a high area on top of a cupboard or a shelf. Any places provided should be located in such a position that the cat is able to get down; it is always easier to climb up. Here are some suggestions for suitable locations.

- Tall scratching posts are available as modular units and they are often floor-to-ceiling structures. Many provide platforms and enclosures for resting and represent challenging climbing frames. If you or a member of the family is a DIY enthusiast in every sense of the word then you can make your own modular centres. Sisal rope can be purchased in pre-cut lengths for wrapping round the up-right pillars.
- Free-standing cupboards and wardrobes have large areas where a cat can rest or hide in a high place. It may be necessary to place furniture nearby to give your cat a halfway platform for ease of access. They are more likely to find this area appealing if you don't signal its presence too strongly or try to put them there yourself. Cats have a naturally suspicious element to their character and over-enthusiasm on your part may be the best way of ensuring they avoid the area at all costs.
- Shelves can be constructed specifically for your cat's use. It is important to provide a non-slip surface as many wooden

shelves are extremely slippery and your cat could end up leaving the shelf sooner than he intended if he takes a giant leap to get there. Existing bookshelves and other shelving can also provide sanctuary if a small area is cleared for your cat's use. Keeping expensive breakable ornaments on shelves or mantelpieces is inadvisable in a household with an indoor cat. Once he gets used to travelling around the home using every surface apart from the floor you will need shares in Superglue.

- Securing a section of close-weave carpet to a wall represents a challenging climbing frame. This can be fixed by attaching double-sided adhesive carpet tape to a clean wall surface. The carpet is then stuck to the back of the tape and wooden batons are positioned at the top and bottom (secured with screws and rawlplugs) for added security. It may be advisable to have shelves or cupboards nearby to enable your cat to come back down without too much trouble.

- A heavy-duty cardboard tube from the inside of a roll of carpet can be utilized indoors. Covered with carpet (inside out), or sisal twine if you are feeling particularly rich (as you will use a great deal), and secured to a wall or ceiling, such a tube is about as near to a tree as your cat will get indoors. It makes a great climbing frame for your cat and will provide hours of exercise.

Private areas

Cats need 'time out' from owners and other cats in the group so there must be a number of places where they can hide without fear of being discovered. They can be under the bed, inside cupboards or wardrobes or behind the sofa. The golden rule is that these places are sacrosanct and a cat should never be disturbed or acknowledged whilst using one.

Beds

There is a wide variety of commercially manufactured beds including everything from a leopard-print beanbag to a wooden four-poster, but cat beds are rarely chosen by the average cat when they can have an alternative such as the owner's bed, a chair or a sofa. Your cat will want to sleep in different places depending on mood or time of day. He will choose quiet places, sunny areas or warm laps in front of the television. A radiator hammock is great in the winter for those cats who have been heat-seeking missiles in a previous life. If your cat reckons 'if it's good enough for you it must be good enough for me' you may want to personalize his area on your bed with a synthetic thermal fleece to contain all the fur and muddy pawprints.

Scratching posts

Cats need to scratch to maintain their claws and mark their territory (see Chapter 5). If alternative provisions are not made then cats will scratch items of furniture. Scratching posts should be as tall as possible to allow the cat to scratch vertically at full stretch. Panels can be attached to walls at the appropriate height if space is at a premium. Some cats prefer to scratch horizontal surfaces so a variety of scratching areas should be provided. Once again, the maxim 'one per cat plus one' will avoid any obvious competition in multi-cat house-holds.

Scent stimulation

Catnip

Two-thirds of cats respond to the smell of the herb catnip (*Nepeta cataria*). Cats will rub their faces on it and even eat it and the resulting effect on their behaviour seems to range from a profound calm to a temporary euphoric state. Some cats can even become quite aggressively excited so I would advise you to offer your cat a small amount first and stand back and see what happens. If your cat responds to catnip it can be used sparingly to provide a fun distraction in all sorts of different ways. Manufacturers of cat products have capitalized on the popularity of catnip and marketed it in varying forms includ-ing packets of loose dried herb, edible treats, aerosol spray, impregnated toys and even soapy bubbles. Probably the most potent is the fresh growing version in your garden although it is very hard to keep a plant to maturity as it is so totally trashed by all the cats in the neighbourhood. If it is grown surrounded by other sturdy plants or through a topiary wire you might get an established plant. The next best alternative is probably the packets of dried catnip sold in most pet shops.

Valerian

This herb has a calming effect on cats and some respond extremely well to valerian tea bags. These can be offered dry to cats (remove any string or staples) and they will rub and roll on them, imitating their response to catnip. Valerian tea bags can be placed in cardboard boxes or food-foraging triangles as previously described to encourage exploration with a good reward at the end. Some cats may be tempted to eat the bags so it is best to provide one under supervision first to monitor the reaction.

Predatory play

Fishing-rod toys are ideal to simulate the movement of prey. These can be highly entertaining for your cat with the minimum of effort on your part. The toy should be agitated in front of the cat (not in a rhythmical swing but with a random movement) in such a way as to allow the cat to catch it from time to time. Cats will often watch avidly and only pounce when the object of their attention is still, so remember to stop the thing moving occasionally. There are a number of good fishing-rod toys on the market but it is best to avoid those with large heavy objects on the end; you don't want to render your cat unconscious with an over-enthusiastic swing. These toys are best stored away when they are not being used because your cat might just get tangled in one if he tries to play with it when you are not there. If you fancy a bit of DIY why don't you try making your own fishing-rod toy? Here are a few suggestions:

- Use a garden bamboo cane for the rod.
- Attach string or fine elastic to the end of the cane using a very sticky adhesive tape, such as carpet- or duct-sealing tape.
- Some cats will like the string on its own without anything attached.
- Have a variety of small items that can be attached at

different times to maintain the novelty of the toy.

- Try a twisted cellophane sweet paper (this looks like a butterfly).
- Try two feathers (this can be made to 'fly' above your cat; very tantalizing).
- Try a small strip of real fur or a small commercially made fur mouse (approximately 2 centimetres long).
- Try four pipe cleaners bent around each other to form a spider.
- Try attaching a thin strip (approximately 1 centimetre wide × 1 metre long) of thick fleece material instead of the string; or attach this to another cane as a separate toy.

Many cats enjoy retrieval games, which can represent an opportunity for social contact as well as play. The elasticated towelling hair bands are just the right size for a cat to pick up and they seem to be very popular. The younger your cat is when you start the better, but even adults can get the idea that the game will continue if they bring the hair band back to you.

Toys

It is also useful to have toys that your cat can play with when he is on his own. Toys soon become predictable and boring if they are allowed to remain motionless in the same place all the time, so think twice before you place a basket of cat toys in the corner of the room. Toys made out of natural fur and feather of a similar size to prey animals are popular but many owners find the use of real fur rather distasteful. It will not encourage your cat to go out and kill anything remotely furry; bear in mind that this is basically what they are designed to do anyway. Many commercially available toys are made from fur that is a by-product of a food source, which may make the non-vegetarians among you feel a little easier about purchasing them. Toys should be stored away in a self-sealed polythene bag with a small pinch of catnip inside if your cat is susceptible to the charms of this amazing herb. A random selection can then be brought out daily to maintain their novelty. Small toys (like fur mice) can even be placed inside the food foraging receptacles for a bit of added interest. A word of warning regarding small fur mice: some have plastic noses and eyes and these are best removed before your cat plays with them.

Here is a selection of rubbish that your cat may find fascinating and won't cost you a penny.

- A scrunched-up piece of paper thrown across the floor (tin foil works just as well).
- A cork (champagne, of course!).
- The plastic seal on the top of a milk container (under supervision only, as this is quite small and could be swallowed).
- Cardboard boxes.
- Paper bags.
- Supermarket carriers (handles removed).
- A walnut (they make a great sound).
- An empty crisp packet tied into a knot.

Over the years I have acquired a selection of tatty objects in my consulting briefcase. These need to be really good toys because I often get one chance to interact with my patients before they lose interest. I have a fishing-rod toy, several pieces of fur, a feather, a few small fur mice and my secret weapon – the Octopuss. Every owner I visit covets this toy so I have detailed below exactly how you can make one for yourself.

When you have finished you may wonder why you bothered because it doesn't look anything like a cat toy. Be patient, follow the instructions carefully, then put it in front of your cat and let him get on with it. If he likes catnip, he'll love the Octopuss.

The Octopuss

- Purchase a pack of loose dry catnip (20 grammes approx.), available in most good pet shops.
- Place half the contents in a small self-sealed polythene bag (13cm × 9cm approx.).
- Pierce the bag four times with a skewer (or similar) to allow the scent of the catnip to escape.
- Cut out two rectangles of material from a towel or face flannel (18cm × 10cm approx.).
- Place the two rectangles together with the back of the material on the outside and sew around three sides (with a margin of 0.5cm), leaving one of the narrower sides open.
- Turn right side out; this forms the body of the Octopuss.
- Place the filled self-seal polythene bag inside the pouch together with a sheet of greaseproof paper (9cm × 16cm approx.).
- Pad the inside of the body with additional strips of towelling so that it fills out slightly without being over-stuffed. Sew up the open side.
- Cut eight strips of the same towelling (5cm × 12cm approx.).
- Fold in half lengthways, with the back of the material on the outside, and sew up the side and the bottom. Turn right side out; these form the legs of the Octopuss.
- Attach the legs by sewing the tops to the bottom of the body in a circle about 0.5cm from the seam.

- Rub the Octopuss with a small pinch of dry catnip and then spray lightly with Feliway spray.
- Place in a self-sealed polythene bag containing a pinch of catnip and allow the smell to infuse the toy.
- After a couple of days offer the Octopuss to your cat for fifteen minutes a day.
- If he likes it he will rub his face on it and grab hold of it and kick it whilst plucking at the towelling with his teeth. Success!
- Store it in the polythene bag with the catnip when not in use.

There are very few cats who don't find this toy irresistible but don't force the issue. It is probably larger than toys your cat has previously played with so let him take his time to discover its charms slowly.

Novel items

Your cat now has a multitude of high resting places, beds, scratching posts and feeding stations but even these, after a while, will be predictable and potentially boring. Whilst a degree of predictability is very reassuring it is still important to challenge your cat with exciting new experiences. Anything that comes in through the front door is worthy of investigation because it will be laden with a host of different aromas that are well worth a good sniff. New items should therefore be brought into the house on a regular basis to challenge the cat's sense of smell and desire to explore novel things. Wood, stone, plants, cardboard boxes, paper bags, etc. can be placed in various locations and left for your cat to explore. It is important that your cat is regularly vaccinated and treated for parasites if items could potentially have been in contact with other cats outside.

Fresh air

Nothing contains more cat information than a good dose of fresh air; watch your cat the next time he steps out of the house and just sits and sniffs for a while. There are a number of secure grills that can be fitted to open windows that will allow fresh air to enter the house without the risk of your cat's falling out or escaping. Challenging smells will be carried in from outside and become a focus of attention for the bored house cat.

In my opinion, it can never be ideal to keep cats exclusively indoors, but there are numerous reasons why some owners feel that it is the only option. An understanding of the cat's need for a stimulating environment will ensure that he remains as happy and healthy as possible. I hope these suggestions have inspired you to create a haven of exciting activities for your house-bound cat.

CHAPTER 12

Common Cat Conundrums

PEOPLE OFTEN WANT TO KNOW WHAT ARE THE MOST frequently asked questions I come across in my work, and although the answers to all the queries below can be found elsewhere in this book, or in *Cat Confidential*, it seems a good idea to bring them all together in one chapter for the sake of everyone who needs an answer *now*. They can read the rest of the books at their leisure later!

How do I know which kitten to choose?

Choosing a kitten should always be a labour of the mind rather than the heart. It is always better to ask questions before you view, since it is easier to reject a litter as unsuitable before you see their little faces. If the answer to any of the following questions is 'No' then it may be better to look for another litter:

- Have the kittens been reared in a domestic home?
- Have the kittens been handled by a number of people from the age of two weeks?
- Can the mother be viewed with the kittens?
- Is the mother friendly and outgoing?
- Have the kittens been examined by a vet and have they been wormed and treated for fleas?

If you are viewing kittens in a rescue centre it is important to find out as much as possible about their background. Do not be tempted by feral kittens that spit and hiss and are reluctant to be handled. They will be very hard work and should only be taken on if the family understands they may never be the ideal friendly pet. I would also recommend you avoid buying kittens from pet shops. Any kitten you choose should fulfil the following criteria:

- Bright eyes with no discharge.
- Clean anus with no sign of diarrhoea.
- Clean ears with no evidence of dark brown wax.
- Shiny coat and no pot-belly (which would indicate a worm burden).

- Alert and interactive.
- Playful with the other kittens in the litter.
- Keen to approach strangers.

Should I get one or two kittens?

Single kittens are advisable as companions for existing cats in the household. If you work during the day then a pair of kittens will be good company for each other when they are growing up, particularly if you plan to keep your cats exclusively indoors. Given the complexities (and my own personal opinions) of multi-cat households I would always recommend one cat in the household with access to outdoors. If the cat needs social interaction with its own species it can achieve this in the territory outside.

How can I introduce a new kitten to my existing cat?

A little bit of extra effort at the beginning can make the difference between a good and a bad relationship in the future. Your cat has a den of his own and the introduction of another, albeit a little kitten, is not necessarily going to make his day. Here are a few suggestions:

- Place a kitten pen* in a room that your existing cat doesn't particularly favour.
- Allow the kitten to exercise within the room when the other cat is not around.
- Open the door to the room whilst the kitten is eating in his pen and place a bowl of tasty food as near to the pen as you

---

* A kitten pen is a large metal cage with a solid floor that is normally used for kittening queens, or cats who need to be confined after surgery. It is quite large with plenty of room for a bed, toys, food, water and a litter tray. They are often available for hire from veterinary practices or you can purchase one from any good pet shop.

can for your existing cat to eat comfortably.

- Reduce the distance between the two cats when they are feeding by small amounts daily.
- Exchange bedding between the two to allow them to become familiar with the other's scent.
- Provide attention to the existing cat but do not exceed the amount that he finds enjoyable.
- Start to place the kitten pen in other rooms of increasing importance so that your existing cat will understand that the kitten has rights of access to all areas.
- Allow several weeks before opening the pen and letting the cats get to know each other.
- Keep a cushion or pillow handy to place between them just in case things do not go according to plan.

How do I introduce a new adult cat to my existing cat?

Confinement in a kitten pen can be quite distressing for an adult cat. I would recommend that the new cat is kept in a single room first rather than a cage. The existing cat should then be introduced gradually by following three basic steps:

Step 1   Scent – your cat should first be aware of the scent of a new cat. Cats have glands around their head that secrete a pheromone that signals a positive message of security and familiarity. The new cat's scent can be collected and deposited in areas where the existing cat is housed and vice versa (see Chapter 2 for the best way to collect this pheromone from your cat).

Step 2   Sight – your cat should then be able to see the new cat before he is able to have physical contact. A wire frame to fit within the door surround can be useful for this purpose.

Step 3   Touch – physical contact can then be established after a reasonable period of time.

Each cat will respond to the scent and sight of a new cat

differently so the period over which the introduction is made will depend on the personalities of the individuals involved. It is often tempting to interfere in their initial interaction but, unless they risk injuring each other, it is usually best to let them sort it out in their own language.

The most difficult cat to introduce is one entering a house where the resident cat has previously lived alone, as the original cat is inclined to resist the concept of sharing. Introducing cat number three, four or five is usually less traumatic, but careful selection is always important (see Chapter 7 for further details).

How do I prepare my cat for long car journeys?

If you have ever been in a car travelling with your cat you will know how distressing is the howling, vomiting, urinating and defecating that accompany many journeys. If you have the opportunity to accustom your cat to car journeys from a very early age then do so – it will pay dividends in the future. My hand-reared cat Annie used to travel to work with me by car when she was only a few weeks old, so that I could give her regular feeds. She is the most relaxed cat imaginable on a trip to the vet because car travel has always been a part of her life. If it is just too late to acclimatize your cat gradually, here are a few suggestions.

- Make sure that you have a secure basket for transportation. Do not be tempted to allow your cat to roam around the car because this can be extremely dangerous.
- Spray the cat basket with synthetic feline facial pheromones (available from your vet) half an hour before the journey. This will have a calming effect.
- Do not spend the whole journey reassuring your cat.

Remember that comforting an inappropriate fear merely reinforces it. Just talk normally to any human companion or keep quiet and ignore the yelling.

- Make sure your basket is lined with polythene just in case your cat has an accident. A kind veterinary nurse will always mop up at the other end but it's worth taking polythene bags, spare bedding and plenty of kitchen paper if your cat is prone to intestinal hurry.
- Some flower essences are good for car journeys, particularly mimulus. A homoeopathic vet will recommend a treatment for car sickness if the problem becomes unusually difficult.
- It is best not to feed your cat before a car journey.
- If you are planning a long journey then your vet may recommend the use of a mild sedative.

My cat hates trips to the vet. What can I do to make them easier?

The best way to tackle this problem is to plan ahead. The cat basket, often a trigger for a disappearing act, should be left out at all times. It should contain a warm and cosy bed and the occasional dry food treat to encourage use and positive associations. Instead of a portent of doom, the basket then becomes a friendly little resting place to curl up in after a hard day. When the day comes to visit the vet it is important not to get distressed yourself. You may think you look normal as you approach your cat to attempt his capture but you probably look criminally insane and definitely not to be trusted. Try to relax. Once he is safely imprisoned in the basket you can follow the advice for stress-free car travel above.

Should I fit a cat flap so my cat can come and go?

Cat flaps are a mixed blessing. The invention was heralded as a great advantage for cats, giving them freedom of choice to be inside or outside when their owners were at work. It seemed like the perfect solution, but maybe we should view these contraptions from a different perspective. Have you ever noticed your cat jump when he hears the sound of the cat flap? It may be his companion coming in after a hard day's hunting but it could also be the nasty tom from next door. Cat flaps let every Tom, Tigger and Sooty in the house and your cat knows it. There are of course magnetic or electronically controlled cat flaps that allow exclusive entry to key-holders only. They require a collar with a rather cumbersome box on it to be worn full-time by the resident cat but they should, in theory, stop intruders. Unfortunately the whole concept of exclusive entry is often difficult to grasp for the average cat and he will probably still sit and guard the flap from time to time and generally believe that it represents a great weakness in the home's defences. This is a terribly difficult decision to make. Some owners allow their cats out when they are at home by providing access via a door or a window. This does restrict your cat's movements during the day but it's better than no outdoor life at all. Other owners with a slightly more relaxed view of life will go to work and lock the doors, leaving the cat in or out depending on where he happens to be at the time. The cat therefore could potentially do its own thing outside during the day (they really are quite used to the outdoor life) and be waiting on the doorstep to greet his owner after a hard day at the office. You could of course take a chance on a cat flap and monitor your cat's emotional state. If you are very lucky there will be no assertive and opportunistic cats in the neighbourhood queuing up to raid the house when your back is turned.

What is better for my cat, home alone or cattery?

This really does depend on your cat and the relationship you have with each other (see Peanut's story in Chapter 10). Most cats enjoy the routine of familiar surroundings and will cope extremely well with visits from neighbours, family or professional house-sitters to provide food and company when you are away. There are some, however, who want you and nobody else and, in your absence, will feel a strong sense of being 'home alone'. These rather dependent little cats will get into a bit of a panic when confronted with the prospect of protecting the house from invasion all by themselves. They may mark important areas like your bed with faeces to warn intruders that there is a combined force to be reckoned with. You will return and be disgusted with Tigger's 'dirty protest' when, all alone, he has gone through seven kinds of hell whilst you have been sunning yourself. Don't forget, cats don't do dirty protests. It's really not their motivation at all. Tigger and cats like him need a gentle introduction to the concept of being without you. This could take the form of getting him used to the idea of being alone by taking day trips or weekends away before embarking on a full-scale holiday. The alternative to care at home would be a visit to a reputable local cattery. If regular cat holidays are introduced from an early age then cattery stays do not have to be traumatic. It is worth doing your homework first to find a suitable establishment and then booking well in advance. Good catteries get very busy! Here are a few tips for finding the right holiday home for your precious puss.

- Visit the cattery during their published opening times or phone first and make a specific appointment. Ask to be shown the facilities and pay particular attention to the security of the premises and the overall cleanliness. A smell of disinfectant could be a warning sign: it might be an attempt to mask a general lack of good hygiene.

- The cat enclosures should be large enough to provide an enclosed sleeping area and an open-air exercise run with barriers between each pen to prevent the airborne spread of contagious diseases.
- Ensure the cattery has isolation facilities for any cats who become ill during their stay. It's important that they ask for your cat's vaccination history, and an enquiry regarding any special dietary or medication requirements will show that they are caring and knowledgeable proprietors.
- The availability of heating and cooling equipment should be checked to ensure that the temperature within the enclosed sleeping areas can be controlled for the comfort of the resident.
- The cattery should be well maintained and the concrete areas free from any green algae stains.
- The food and water bowls provided for the cats should be clean and the litter facilities regularly checked for soiling.
- A good cattery will allow you to take some of your cat's familiar toys, bowls, trays and bedding if you wish.
- When you are viewing the premises it is also advisable to check out the residents. Do they look relatively content? This is often difficult to assess but check whether they are active and inquisitive and have good appetites.
- Recommendation from the Feline Advisory Bureau (a cat welfare charity) is always a sign that the cattery you are viewing meets the highest standards.

Do deterrents work if my cat is soiling in the house?

I have been to many homes over the years where I have had to walk through a veritable assault course of tin foil, pepper, pine cones and citrus peel. If these deterrents actually worked you could argue that the owners wouldn't need the services of a cat behaviour counsellor in the first place. Such smelly or noisy objects may deter a cat from soiling in one specific location but never address the underlying cause of the problem. You might just be encouraging your cat to find several other sites to soil. Some well-meaning people still recommend that you rub your cat's nose in any urine or faeces that he may have passed on the floor. PLEASE DON'T DO THIS! It will make your cat even unhappier, he will not have a clue why you did it and you may never be trusted again.

How do I stop my cat catching birds?

Not all cats are adept at hunting birds, but if your cat has this special skill it can be extremely distressing for bird lovers. It is important to keep reminding yourself that your cat is a carnivore and the whole purpose of their existence is to catch and consume enough prey to stay alive. We may provide them with all the food they need but this doesn't stop their innate drive to catch small things that move quickly. Here are a few suggestions.

- Try adding two small bells to your cat's collar. These may knock together and alert any potential prey to the danger. Many cats get wise to this very quickly and hold their neck still but it's worth a try.
- Even if you are a bird lover it is probably best not to encourage birds into your garden with feeders and bird tables. If you feel compelled to do so then it is advisable to make the stand of any table as tall as possible.
- Do not give your cat more food to stop him hunting. The desire to hunt is not triggered by hunger.

My cat's a bully to other cats in the neighbourhood. What should I do?

Some cats have a Jekyll and Hyde quality. They can be sweetness and light to you and the family but Attila the Hun to the neighbourhood moggies. You may be blissfully unaware of this until, quite by chance, you discover that several of the cats in the area have been systematically beaten up and the perpetrator is your beloved baby. Situations of this kind can cause a great deal of conflict between neighbours. Cats can inflict costly damage on others and hefty vet bills on a regular basis are no joke. Owners are particularly upset if your cat enters their home to carry out the act of violence. These situations involve a great deal of tact and diplomacy; I used to try to help by acting as arbitrator but experience has taught me not to get involved any more. Cats are territorial creatures and it is unreasonable to expect disputes between cats to be resolved without aggression. It is natural behaviour; unfortunately some cats behave more 'naturally' than others. Every owner has a duty of care to protect their own cats if they can't stand up for themselves. Territorial aggression is all a matter of degree. I wonder if anyone will ever have to stand up in court and argue what level of aggression is natural for the species. I'd be in the gallery for that one.

Here are a few suggestions for you to deal with this delicate matter.

- Encourage your cat to stay indoors at night. Most fighting occurs during the hours of darkness. Feeding him a tasty late-night treat may be a sufficient incentive to tempt him in at a certain time.
- Inform the neighbours that your cat is confined at night so that they know when their own cats are safe. If your cat is

nocturnal in his habits it may be useful to shut the cat flap during the day and inform your neighbours accordingly.

- Ensure there are sufficient warm beds around the house to give your cat every opportunity for relaxation in a comfortable setting.
- Provide plenty of stimulation indoors (active play sessions, etc.) to use up energy.
- Suggest the neighbours fit exclusive-entry-system magnetic cat flaps and ensure that several cats from different households do not have the same 'keys' on their collars.
- Territorial cats should have a couple of bells attached to their collars so neighbours and their cats can hear them coming.
- Neighbours should be encouraged to keep a water pistol by the back door. One squirt and the element of surprise may deter the less determined cat from entering a neighbour's home.
- Always appear to be doing your very best to resolve the problem. After all, it could have been you and your cat on the receiving end.

My cat's being bullied in the neighbourhood. What should I do?

You see, I told you it could be you! Everything relating to the previous question applies to you also. You have a duty of care to protect your cat if he cannot defend himself. The cat flap may need to go (if you have one) and a new regime started of access outdoors only when you are at home. Your cat may feel more secure in these circumstances. If the territorial cat does ever get access into your home it is essential to separate the antagonists using a pillow or cushion. If you use your body parts then the cat will turn on you too. Speak tactfully to your neighbour and see if any of the suggestions made above could be implemented. Sadly, you are just unfortunate to have a timid cat who lives so close to a lean mean fighting machine. If you are very lucky the aggressor will move house.

How can I make moving house easy with my cat?
Moving house can be traumatic enough without considering the needs of our pets. A cat's environment is an extremely important element in his life and a change to a completely different home can be stressful. The following suggestions may be useful.

- Walnut is a useful flower essence to give to your cat to help him adjust to new surroundings (check with your vet first).
- On the day of the move shut your cat into one room that has been cleared of all large furniture. It may be worth using one that contains a fitted wardrobe or cupboard where he can hide if he gets anxious. Leave food, water, bedding, litter tray, familiar cat toys and scratching posts.
- Make sure the removal team know that this door must be

kept shut to avoid your cat's going missing at the last moment. Once the removal van has gone you can take your cat to the new house in your car.

- In the new home it would be useful to have one bedroom's contents unloaded first so that there is at least one room containing your furniture. Plug a diffuser in a floor-level socket that emits synthetic feline facial pheromones (available from your vet). This will make your cat feel that he is in familiar surroundings. Place your cat in this room with food, water, bedding, litter tray and all his accessories and leave him quietly to explore whilst you supervise the rest of the move.
- Your cat can be allowed into the rest of the house once it is quiet and all the furniture is in place.
- If your cat is nervous you may want to allow him to explore smaller areas at a time. Be relaxed and don't reinforce anxious behaviour by reassuring him.
- If your cat is confident and loves to be outdoors you may find it difficult to confine him for any length of time. The recommended period is two to three weeks but, in practice, this is often impossible. If you have moved relatively close to your old home be warned that some cats will find their way back.
- When you first allow your cat to explore outside it is best to choose a weekend when you are at home and let him out just before a mealtime.
- Ensure your cat is microchipped and has a collar showing his address or telephone number.
- Your cat will now have to establish rights in his territory so you may find he gets involved in more altercations with the neighbours' cats than usual. This unfortunately is all part of a move to a new area. No welcoming barbecues for your cat, I'm afraid.

What do I do when my cat keeps returning to my previous home?

Every now and then you will find a cat who is more attached to his environment than he is to his owner. If that cat moves within a relatively short distance of his beloved hunting ground then he will probably try to return. Main roads, railway lines and rivers will not deter him. I often speak to owners who are experiencing the stress of constantly returning to their old homes to find their cat looking confused in the back garden, wondering why he can't get indoors. There are some owners who accept the fact that their cat is happier in his old stomping ground. The new occupier of the house or a willing neighbour then becomes the custodian of the cat and he carries on virtually as if nothing has happened. This is tough for you but the alternative is to keep returning and collecting him and bringing him back. Plenty of palatable treat food and entertaining games will show him what an excellent place his new home is. Unfortunately it is inappropriate to confine him indoors because when he does eventually go outside (you can't keep him in for ever) you will find it harder to get him back the next time. I gave advice to a lady once who persevered with her cat for six months, collecting him and bringing him back two or three times a week. He has now finally got the message.

My cat is grieving for his dead companion. Should I get another one?

Back in 1995 I conducted a survey relating to the changing behaviour of the elderly cat. It was impossible not to touch on the whole concept of feline grief. Do cats mourn the death of a companion? Many of the 1,236 owners surveyed sent accompanying letters with their completed questionnaires and these were a great source of information. By far the largest number of letters I received related to the reactions of their elderly pets to the death of a cat friend. Nearly half of the cats surveyed had outlived another and 60 per cent of those showed some visible reaction to the loss. The Siamese, Burmese and Birmans appeared to be particularly affected. Almost all the reactions reported included searching and calling. Some told of their cats becoming more affectionate and demanding; some even said the surviving cat improved tremendously and appeared more content after the loss of the other cat.

I do not feel this 'grief' is behaviour unique to old age. Similar instances have been reported in younger cats. The relevance of age is that the companions have often been together for a very long time and the desire for routine and lack of change appears to be heightened in the elderly. The loss of a long-term friend creates a profound difference in the house-hold – grieving humans, changes of routine and the absence of a familiar part of the family unit probably causes the distressed calls and searching to try to return things to normal. The intro-duction of a kitten can sometimes stop the unsettling behaviour by occupying the mind with a new source of company. There is also the other side to the coin: those owners who reported the remaining cat 'blossomed' on the demise of the other. It appears that passive oppression between cats may only become apparent when the assertive one is no longer there.

If your cat has lost his companion and appears to be griev-ing then it is important to maintain normal routines as much as you can. Don't worry if he goes off his food for a couple of days; appetite usually returns fairly quickly. Honeysuckle is an appropriate flower essence to give to cats who have lost a companion but please refer to your vet first.

Before you rush out and buy another kitten to 'replace' the departed pussy it's probably worth waiting to see how the grieving cat progresses. After all, if we had lost a partner would we really appreciate a well-meaning friend's bringing a stranger into our house as a replacement? If your cat is grieving for that one unique character and companion it would be better to allow the process to develop naturally and wait for time to heal. If your cat 'blossoms' you know that you have made the right decision. If he continues to mope around the house and look generally lost then you may want to give that kitten a try.

I'm pregnant; will my cat accept the baby?

Many cats are incredibly unperturbed by the arrival of a new baby and will merely avoid it when it starts screaming. However, if your cat has always been your baby before the arrival of the real thing then you may have problems. It is important to plan ahead and, during the pregnancy, gradually adopt the lifestyle changes that are almost inevitable when the baby arrives.

- Make the decision about where the baby is going to sleep at night. If the cot is going in your bedroom then it's important to start denying access to your cat early in the pregnancy.
- If your cat is used to having your undivided attention it is important to gradually withdraw from him during the pregnancy.
- Your cat should be encouraged to indulge in activity outside your relationship, including play, exploration of novel items, food foraging and other indoor pastimes.
- Encourage friends with babies and small children to visit so that he gets used to the presence of children.
- Do not comfort him if he appears frightened of them.

Remember, this is an inappropriate fear that should not be reinforced.

- Ensure there are plenty of high resting places where your cat can retreat away from baby when he or she starts to crawl.
- Introduce all the baby accessories such as buggies, cots and changing mats over a period of time to avoid a sudden influx of challenging smells and objects.

My cat's on a diet but he's not losing weight. What should I do?

Obesity in cats is becoming a big issue. It can lead to heart disease, diabetes and joint problems. Most modern veterinary practices now have weight clinics run by veterinary nurses where your cat's target weight and feeding regime is discussed and monitored. However, as anyone who has been on a diet themselves will know, reducing the amount of food given doesn't work on its own. Exercise is the key to sustainable weight loss. If your cat is on a diet that you have created yourself merely by reducing the amount of food given then it is likely you need help. Your veterinary surgery will provide an alternative diet food that will allow weight loss to be more easily achieved. This should be combined with increased activity in the home to encourage movement and aerobic exercise. It may be useful to place the food on a high surface so that your cat has to climb to get his meals. A feather on the end of a fishing-rod toy will animate even the most inert cat into feverish activity. Providing your cat isn't snacking with the neighbours he will soon be slim again and much healthier.

Can I take my cat for walks on a lead?

Daily walks can be stimulating and enjoyable for some house

cats although a lead and collar is not the safest or most comfortable option. Specially designed harnesses can be purchased that fit securely round your cat's chest and avoid the risk of the cat's hurting his neck or slipping his collar when you are outside. In my experience breeds such as Maine Coon, Siamese, Burmese and Bengal thoroughly enjoy walking on a harness but any cat, providing they are introduced to the concept early, can get some benefit. Here are a few recommendations if you think this may be an option for you and your indoor cat.

- Ensure the harness is designed for use with cats and is fitted correctly.
- Start early and get your new kitten used to wearing a harness.
- Attach the harness initially without the lead.
- Reward your kitten and have a game when he first wears the harness to distract him.
- Attach the harness for short periods initially and increase the time it's worn gradually.
- Always reward your kitten during periods of harness training with food or play.
- Never leave your kitten or cat unsupervised whilst he's wearing a harness.
- Bear in mind the harness will need adjusting or replacing as your kitten grows.
- The lead can be attached and allowed to trail once your kitten gets used to wearing a harness.
- Gradually get your kitten used to your holding the lead and following him around.
- Now you are ready to try the process outdoors.
- Ensure the first visit outside is in a safe and quiet area.
- Gradually increase the time spent outside.

- Avoid walking on busy roads or in parks with dogs. Cats can become easily frightened and their instinct will be to escape.

Is an enclosure in the garden a good thing for an indoor cat?
Keeping cats exclusively indoors is never an ideal option (see Chapter 11) so if there is an opportunity to increase their stimulation by providing a secure pen outside it would certainly be an advantage. Here are a few tips that may help you design your structure.

- A concrete or patio slab base will enable the enclosure to be used all year round.
- The ideal structure is wooden-framed (weatherproofed with animal-friendly preservative) with wire mesh attached to the outside.
- If the enclosure is south-facing it is important to provide areas within it where your cat will be protected from direct sunlight.
- The roof should be covered with a waterproof material to guard against the worst of the weather.
- The wall of the house can form part of the structure, with or without a cat flap access.
- A door should be included in the structure for ease of access and cleaning.
- The structure doesn't have to be large but the full space can be used by including wooden platforms and shelves.
- Shelves can be accessed from logs or ramps made of wood.
- Include pots of grasses or other cat-friendly plants inside the enclosure.
- Include a water feature if there is sufficient room; if not, a water bowl should always be available.

- An outdoor covered litter tray may also be provided, particularly if your cat does not have freedom of access into and out of the enclosure.

Epilogue

WHILST I HAVE BEEN WRITING THIS BOOK I HAVE CONTINUED to see as many patients and their owners as possible to help them deal with a variety of problems on a one-to-one basis. It can safely be said that there is no substitute, when things really get bad, for bringing in an expert who understands what the family as a whole is going through. My clients often tell me, when their two-month therapy programme is completed, how reassuring it was to have someone at the other end of a phone who could give advice and encouragement every step of the way. I don't think I will ever stop doing the actual job of cat behaviour counselling because I find it incredibly satisfying to be able to restore harmony to cat-loving homes. My biggest frustration has always been that I cannot see as many patients as I would like. *Cat Detective* may just be able to solve that problem. Within this book I have given you a little insight into the workings of cat behaviour therapy. Many of the principles could easily be applied by you, the owner.

Unfortunately, as I have suggested throughout this book, there are some occasions when problems just cannot be fixed. This is by far the most distressing element of the work and something that every cat owner needs to consider. What would you do if you were advised (and then grew to understand yourself) that the best outcome for your cats would be for them to be permanently separated? How would you feel if, despite all your love and attention, your only cat was extremely unhappy living with you and needed to find another home in order to recover from a stress-related illness? This is an extraordinarily bitter pill to swallow and it is almost impossible not to take it personally or presume that you are a hopeless individual because even your cat hates you. I made the decision, when I moved to Kent to pursue my work, that my beloved cats would remain in their wonderful home in Cornwall. Not a day goes by when I do not miss them but I have never regretted that decision. I have seen many households over the years where love has been expressed in a very selfish guise. Owners maintain difficult and highly distressing multi-cat relationships because they fool themselves into believing that their cats would miss them if they were re-homed, or that their decision to re-home them would be viewed by others as neglect. If I manage to get one point across within this whole book I would like it to be that sometimes you need to love a cat enough to let it go. Many of my clients have made this incredibly brave decision and, having seen the results, have been entirely satisfied that it was the right thing to do.

I cannot finish this book without expressing gratitude to all the clients I have seen over the years. Without exception, they have been wonderful people with an incredible love for their pets. I would not have been able to learn as much as I have about the art of cat behaviour counselling if they hadn't followed my programmes so diligently and reported so comprehensively. I remain in contact with many of them and it is always a delight to hear from them with a progress report on their now happy cats. I fully intend to keep learning how to play cat detective for many years to come.

Useful addresses

Feline Advisory Bureau
Taeselbury
High Street
Tisbury
Wiltshire SP3 6LD

Cats Protection
17 Kings Road
Horsham
West Sussex RH13 5PP

Association of Pet Behaviour
Counsellors
PO Box 46
Worcester WR8 9YS

GCCF (Governing Council
of the Cat Fancy)
4–6 Penel Orlieu
Bridgwater
Somerset TA6 3PG

Index

The relationship survey

Vicky Halls is currently researching the special relationship that individuals have with their cats for her next book, *Cat Kin*. If you would like to participate, please complete this questionnaire and send it to: Vicky Halls, The Relationship Survey, PO Box 454, Rochester ME1 1WZ.

About you:

Tick the relevant boxes:

Age: 18–30 ☐   31–40 ☐   41–50 ☐   51–60 ☐   61+ ☐

Sex: M ☐   F ☐

Marital status: married ☐   single (never married) ☐   divorced/separated ☐   widowed ☐

No. of children living at home: none ☐   1 ☐   2 ☐   3+ ☐

Working: Full-time ☐   part-time ☐   retired ☐   not working ☐   unable to work ☐

No. of cats in household: 1 ☐   2 ☐   3 ☐   4 ☐   5 ☐   6 ☐   7+ ☐

Any other pets: Yes ☐   No ☐

About your cat:

Name of your cat (*if you have more than one cat then only include details of the individual with whom you have the best relationship*) ....

Age of cat: up to 2 yrs ☐   3–5 yrs ☐   6–8 yrs ☐   9–11 yrs ☐   12+ yrs ☐

Sex: Male (neutered) ☐   Female (spayed) ☐   Male (entire) ☐   Female (entire) ☐

Pedigree/breed (*please specify*) ....

Indoor/outdoor:

My cat has unlimited access outdoors ☐

My cat has access outdoors but I shut him/her in at night ☐

My cat has restricted access outside under supervision ☐

My cat is taken out for walks on a harness and lead ☐

My cat is kept indoors but has an outdoor pen ☐

My cat has access outdoors but doesn't go out ☐

My cat is kept exclusively indoors ☐

What kind of relationship do you have with your cat?

My cat is a pet ☐

My cat is a family member ☐

My cat is like a child ☐

My cat is a companion ☐

I find it hard to explain my relationship with my cat ☐

Other ....

Please read the following statements carefully in each section and indicate in the relevant boxes whether you 'agree', 'disagree' or are 'not sure'.

Section 1. Safety

	AGREE	DISAGREE	NOT SURE
I worry that my cat will come to harm when he/she is outside	☐	☐	☐
I worry about the dangers of traffic	☐	☐	☐
I worry that my cat may go missing	☐	☐	☐
I worry about my cat fighting with others and getting injured	☐	☐	☐
I worry that someone may steal him/her	☐	☐	☐
I like to know where he/she is at any given time	☐	☐	☐

Section 2. Interaction

	AGREE	DISAGREE	NOT SURE
I approach my cat for interaction more than he/she approaches me	☐	☐	☐
My cat approaches me for interaction more than I approach him/her	☐	☐	☐
My cat and I approach each other an equal number of times throughout the day	☐	☐	☐
Sometimes my cat will ignore me or seem uninterested when I approach	☐	☐	☐
I will respond to my cat at any time of the day or night	☐	☐	☐
I always respond to my cat's approaches	☐	☐	☐
I will check to see where my cat is if I haven't seen him/her for a while	☐	☐	☐
I will occasionally wake my cat up to stroke him/her	☐	☐	☐
I will seek my cat out if he/she is hiding	☐	☐	☐
My cat will seek me out if I go into another room	☐	☐	☐
My cat is by my side most of the time I am home	☐	☐	☐
I can't sit down without my cat jumping on my lap	☐	☐	☐
I talk to my cat occasionally at feeding time	☐	☐	☐
I talk to my cat all the time	☐	☐	☐
I hardly ever talk to my cat	☐	☐	☐
My cat is very vocal and 'talks' back	☐	☐	☐

Section 3. Bedtime

	AGREE	DISAGREE	NOT SURE
My cat sleeps with me on the bed	☐	☐	☐
I shut my cat out of the bedroom at night	☐	☐	☐
My cat is given the choice but chooses to sleep elsewhere at night	☐	☐	☐
My partner doesn't like my cat sleeping with us	☐	☐	☐
I don't want my cat sleeping with me	☐	☐	☐

Section 4. Personality

	AGREE	DISAGREE	NOT SURE
My cat is loving and attentive	☐	☐	☐
My cat is sociable with everyone	☐	☐	☐
I am the only person my cat wants to be with	☐	☐	☐
My cat is very confident	☐	☐	☐
My cat is self-reliant	☐	☐	☐
My cat is needy	☐	☐	☐
My cat can be aloof	☐	☐	☐
My cat is very timid	☐	☐	☐

Section 5. Relationship

	AGREE	DISAGREE	NOT SURE
My cat knows what I am thinking	☐	☐	☐
My cat understands when I am ill	☐	☐	☐
My cat understands when I am depressed	☐	☐	☐
My cat responds adversely when I am stressed	☐	☐	☐
My cat would miss me terribly if I went away	☐	☐	☐
My cat prefers my company to anyone else's	☐	☐	☐
My cat understands what I am saying	☐	☐	☐
My cat has a sense of humour	☐	☐	☐
My cat would struggle to cope without me	☐	☐	☐
I would struggle to cope without my cat	☐	☐	☐
I wouldn't re-home my cat even if I knew he/she would be happier somewhere else	☐	☐	☐
Nobody would care for my cat like I do	☐	☐	☐

Section 6. Love and support

	AGREE	DISAGREE	NOT SURE
My cat has given me comfort at a particularly difficult time in my life	☐	☐	☐
My cat has supported me through bereavement	☐	☐	☐
My cat has supported me through a relationship breakdown	☐	☐	☐
My cat has supported me through illness	☐	☐	☐
My cat feels unconditional love for me	☐	☐	☐
My cat feels cupboard love for me	☐	☐	☐
My cat's love is conditional on the way I treat him/her	☐	☐	☐
My cat doesn't love me	☐	☐	☐
My cat's love is very different from the expression of human love	☐	☐	☐
My cat seems to love me sometimes but not others	☐	☐	☐
If I don't do what my cat wants he/she won't love me any more	☐	☐	☐
My cat will still love me even if I reject him/her occasionally	☐	☐	☐

Section 7. Behaviour problems

	AGREE	DISAGREE	NOT SURE
My cat has shown aggression towards another cat in the household	☐	☐	☐
My cat has shown aggression towards me or another person	☐	☐	☐
My cat has soiled in the house	☐	☐	☐
My cat has sprayed urine in the house	☐	☐	☐
My cat has been treated for a behaviour problem	☐	☐	☐

How would you summarize your feelings for your cat in fifty words or less?

....

....

....

....

....